BAKELITE

RADIOS

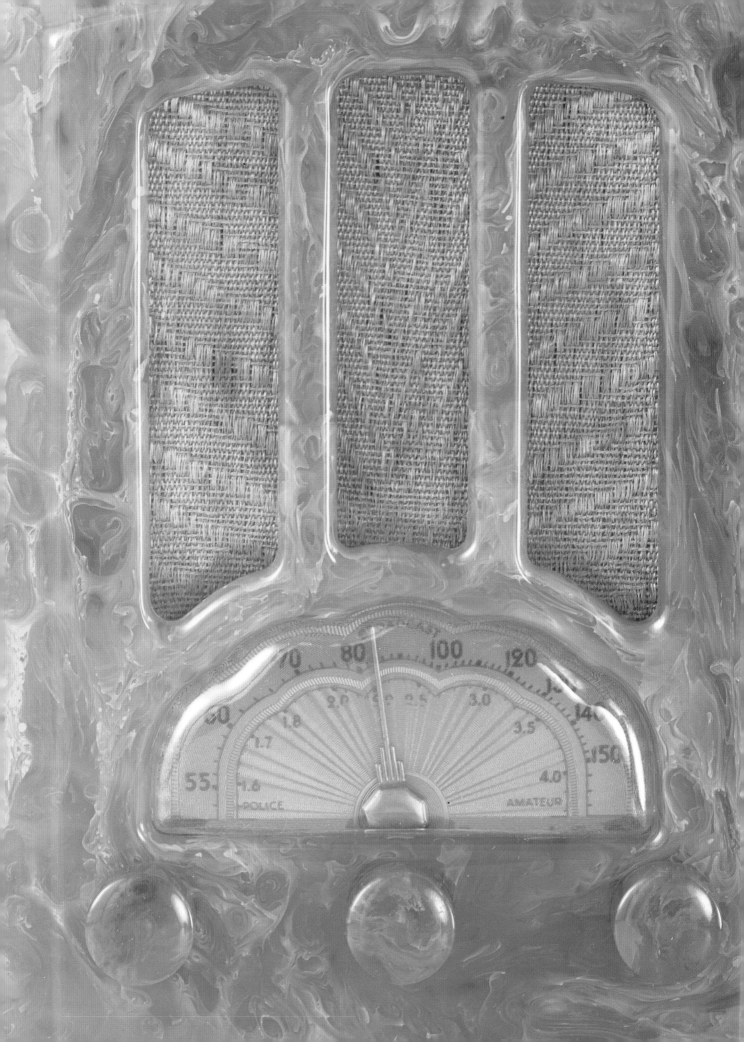

BAKELITE

RADIOS

A FULLY ILLUSTRATED GUIDE
FOR THE BAKELITE RADIO
ENTHUSIAST

ROBERT HAWES
IN COLLABORATION WITH GAD SASSOWER

CHARTWELL
BOOKS, INC.

FOR BHARAT GOSWAMI

"Things are seldom what they seem" – W. S. Gilbert, HMS Pinafore, 1878.

A QUINTET BOOK

Published by Chartwell Books
114 Northfield Avenue,
Edison, New Jersey 08837

This edition produced for sale in the U.S.A., its
territories and dependencies only.

ISBN 0–7858–0389–0

This book was designed and produced by
Quintet Publishing Limited
6 Blundell Street
London N7 9BH

Creative Director: Richard Dewing
Designer: Peter Laws
Project Editor Alison Bravington
Editor: Sean Connolly
Photographer: Nick Bailey

Typeset in Great Britain by
Central Southern Typesetters, Eastbourne
Manufactured in Hong Kong
by Regent Publishing Services Ltd
Printed in China by Leefung-Asco Printers Ltd

ACKNOWLEDGMENTS

Robert Hawes would like specially to acknowledge the help and advice of his researcher Bharat Goswami. He would
also like to thank the following for their valuable assistance: Dr Paul Barnett; Ralph Barrett; Peter Bridgman –
Curator of Communications, Science Museum, London; Lesley Butterworth – Education Officer,
Design Museum, London; Bill Caten; William Coyler-Commerford; Patrick Cook; Arnaud and
Maaiki Cramwinckel; Eryl Davies of the Science Museum, London; John Dittrich and Sony UK;
Ernst Erb; Mary Ann Francis; Philip Graham and the late Stephen Boyd of Philip Graham Gallery London;
Ray Herbert; François Humery; Christopher Jackson for permission to use drawings by his late father
Norman Jackson; Derek Johnson; Juan Julia Enrich; Sylvia Katz; Eliot Levin; Kevin Lowry-Mullins;
Dr Barty Mukhopadhyay; Jim and Barbara Rankin; Roy Rodwell and GEC-Marconi; Wendy Simpson;
Alan Stewart; Dr Rudiger Walz; Eric Westman; Alex Woolliams.

Gad Sassower would like to thank: Keith Banham; Carl Glover; Clive Mason; James Meehan;
Terence Pearson; Patrick Rochette; Patrick Rylands; Philip Rose; Frank Siciliano; John Sideli;
Enrico Tedeschi; Ray Whitcombe; Barry Wilson.

CONTENTS

THE BEGINNINGS OF RADIO 7

Who invented radio?; The Marconi story; The heart of the radio;

The first home radio; Crystal sets – and how they worked;

Aesthetic development; The evolution of the radio

A SHORT HISTORY OF THE RADIO CABINET 27

Small firms; The domestication of radio; The Golden Age of the cabinet;

The evolution of the radio cabinet; Industrial development and

mass production; Cabinet materials; The need for plastics

A SHORT HISTORY OF BAKELITE AND OTHER PLASTICS 43

Bakelite and radio; British round Ekco radios;

Early bakelite manufacturing; The aesthetics of plastics

THE PRESERVATION AND CARE OF BAKELITE RADIOS 59

Sources of supply; Buying bakelite radios: what to look for; Inside radios;

Protecting your collection; Beware! Don't plug it in; Will your radio work?;

Care and repair; Postscript

COLLECTOR'S DIRECTORY 69

Bibliography 127

Index 128

Picture Credits 128

CHAPTER ONE

The Beginnings of
Radio

Radio is now a hundred years old. It began as a scientific toy in the days of Queen Victoria and entered the home as the first really sophisticated piece of domestic electronic equipment: a machine for providing information and entertainment. It developed remarkably rapidly into the world's most important and enduring global communications system and was to found our Age of Electronics, leading to the invention of television, the computer, and all the other marvels of communications technology which continue to change our lives in profound ways. Yet despite our cult of the New, our heritage of antique artifacts continues to fascinate us. Among the most interesting and collectible items from our past are those from the scrap heap of old technology, particularly the things which have the visual appeal of peculiarity, quaintness, and perceived beauty – like the vintage bakelite radios of the Thirties, which can be restored and can still produce authentic sounds from the Golden Age of Radio.

That Golden Age has gone for ever; the combination of circumstances which produced it can never be repeated. It was a celebration of the last stage of perfection of a "steam-radio" technology that was rendered obsolete by the transistor and miniaturized technology which was developed after World War II. It was also a celebration of the art of the cabinet designer and the craftsman of a never-to-be-forgotten period when science and art seemed magically united. The real radio that was has gone out of production altogether.

Extreme miniaturization of the "works" has made cabinets entirely obsolete, except as tiny plastic packages as functional as soap-powder packs. Finally, even these remains have disappeared into the visually

RIGHT A Canadian Crosley of 1953, model D-25, made of bakelite sprayed with white paint.

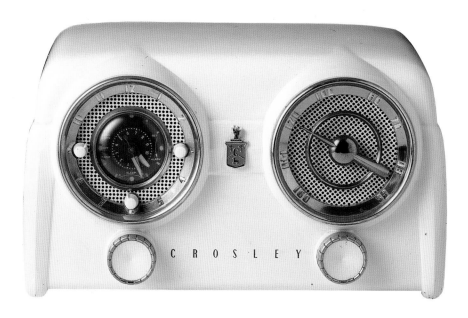

RIGHT An Australian Radiolette, typical of the skyscraper style of the Thirties.

boring black boxes of hi-fi equipment with their multiplicity of "pretend" controls and winking lights, more to do with "hi-tech" fashion than with true functionality.

Like the radio designers of the Golden Age, today's marketing men are still selling us a dream. It used to be a romantic dream, projecting onto the dark vista of the great Depression. The dream was a fantasy image of life made into a dazzling, jazzy, Hollywood movie-star world, full of streamlined cars and labor-saving machines, which were made possible by the emancipating marvels of modern science. Today, consumerism sells us a dream of even more marvelous science and technology. But along with its benefits, this new consumerism has also created a throwaway attitude. As real radiophiles, our best line of resistance is in our own brand of recycling.

Old radios, for a long time considered too recent to be worthy of preservation, are now being accepted as serious objects of study for historians who regard both the technological and social aspects of radio as important. Such artifacts tell us more than their face value. They remind us of our love-hate relationship with technology. They act as icons of our age and symbols of ourselves, reflecting the way we live and how we continuously reconstruct our concepts of our world. A broadly based study of such ubiquitous objects as the radio can reveal important insights which would be invisible if they were considered merely as pieces of antique technology.

In our postmodern Age of Uncertainty, in which the benefits and discontents of Modernity and the rapid modernization of society in the Machine Age are being

ABOVE This Art Nouveau cabinet with spiralling knobs made by General Electric in 1932 was the first bakelite radio manufactured in Australia.

WHO INVENTED RADIO?

Almost anyone would answer: "Marconi, of course." But nobody really invented the electromagnetic waves that we call radio: they were there all the time in Nature! The question of who invented the piece of electronic apparatus we call "the radio" – the receiver and the transmitter – is quite another matter. Until the end of the nineteenth century nobody really had an idea of constructing a "transmitter" to generate and send out electromagnetic waves as a way of communicating through space. Even if such a device had been produced, it would have been no use because no form of corresponding "receiver" had been devised either. Useless natural and manmade interference was the only kind of broadcasting On the Air at that time, but certain pressures led to it being seen as a possible method of communication.

The development of civilization, the march of war, and the spread of commerce increasingly engendered a need for more rapid communication. By the nineteenth century this communication had developed from the very basic – the foot messenger, the horseman, or beacons lit on high ground – to more advanced systems, such as visible signaling by codes using flags and mechanical devices.

Then, at last, science produced the marvel of the telegraph: a system of wires on poles, carrying coded electrical pulses known as "Morse" to be decoded by the listener, then later by electromechanical printing devices. It was an important breakthrough, enabling communication over long distances overland and eventually between continents by means of undersea cables. This revolutionized communications but had many drawbacks, being expensive to set up, labor-intensive, inflexible, and only really suitable for private person-to-person contacts.

With the replacement of the telegraph instrument by the microphone and earphone arrangement of the telephone, the same wires used for the telegraph system were able to be used for voice-communication over considerable distances. This was another great development in communications, but it had the same inherent problems. It still used the old system of connecting people by means of wires, which had serious limitations. Each subscriber needed a separate

reassessed, people seem to crave reminders of a past which have been romanticized by the myth-makers into "The Good Old Days." In those days, life seemed more natural, slower and simpler, and objects were solid, well-made, long-lasting, and beautiful. This is a fantasy nostalgia, and it fails to recall the downside of past ages which could remind us of present woes such as mass-unemployment and poverty.

Similar distortions are at the heart of the way people treat the history of radio. In this case, though, hindsight gives a nationalistic overlay to history. The reality is different. At its beginning, as more-or-less "pure" scientific experimentation, radio was an inter-national affair in which important discoveries and inventions came from all over the world. However, as soon as it acquired economic, social, cultural, military, and political connotations, it became the object of fiercely nationalistic arguments. Various nations claimed the honor of its invention, the rights to its commercial exploitation, and the authority to control it.

RIGHT Guglielmo Marconi, aged 21, photographed in 1896 when he took his "telegraphy without wires" apparatus to London to demonstrate experiments he had been carrying out at his parents' home in Italy a year before. He soon set up the world's first wireless factory, founding a worldwide electronics industry which was to affect the lives of everyone profoundly.

line connected to an ever more complex and developing system of dialing, switching, and ringing equipment. Moreover, it was useless for people situated in places which could not be wired into the system. The most notable examples were ships, which were cut off from all contacts when out of sight or because of distance or bad visibility – a problem later shared by aircraft. Here, good communications could be a matter of life or death – it was in these cases that radio communication literally came to the rescue, very much due to the work of Marconi, although he was not the first to think of the idea.

The specific history of radio can be said to have started when the British physicist James Clerk Maxwell suggested the existence of electromagnetic (radio) waves in 1864, but it was not until more than two decades later that the German professor Heinrich Hertz actually produced such waves by devising an elegantly simple apparatus. He used a sparking-coil connected to a dipole with metal plates at the ends, which acted as a transmitter with which he sent electromagnetic waves a short distance across his laboratory to a wire-loop aerial acting as the receiver.

All he actually transmitted was a single spark – but he was able at will to make it leap invisibly through empty space as if by magic from one piece of apparatus to the other. It was an event which was to revolutionize the world of communications. News of Hertz's demonstration and confirmation of the existence of Maxwell's electromagnetic waves soon reached the scientific world where other pioneers in a number of countries developed different kinds of transmitting and receiving apparatus. One of them was the British experimenter Oliver Lodge, who gave a demonstration of very short range radio in 1894 and suggested that it might serve as a means of communication without intervening wires.

What the 19-year-old Marconi did in Italy soon afterward, was to bring together various pieces of apparatus already devised by other researchers, employing well-known principles to demonstrate a practical system of transmission and reception. This revolutionary system of communication immediately overcame the most serious limitation of its well-established competitor, the telegraph: the need for wires. The new system was "wire-less," with nothing connecting transmitter and receiver. The first stage was "wireless telegraphy," which used Morse code; the next was "wireless telephony." Later, the popular term describing both systems became simply "wireless."

It would be wrong to think that every radio innovator became a household name, as Marconi did. Consider the story of the man who actually built the first practical radio transmitter and receiver, and demonstrated it as early as 1879 – some nine years before the laboratory experiment of Hertz and when Marconi was only five years old! This forgotten man of radio, cheated of his rightful place in history because nobody recognized the marvelous thing he had demonstrated, was London-born Professor David Hughes.

Like many learned gentlemen of his day, Hughes – actually a professor of music – was a keen home experimenter who was specially interested in electrical things and who had earlier produced several successful inventions including a microphone – still an essential device for broadcasting. He had a little laboratory at his home in Great Portland Street, London, which, oddly enough, is on almost exactly the same spot that the British Broadcasting Corporation's Broadcasting House was to be built some fifty years later.

Working one day on a prototype electrical measuring device, without any idea that he could be generating any kind of waves in space, he happened on his discovery by accident. He discovered that he could transmit a spark from one coil of his apparatus to another although they were not electrically connected. Puzzled and excited, he built a crude automatic transmitter from odds and ends: bits of wood, wire, string, cork, sealing wax, old clock parts, and a battery. Next, he made an even cruder receiver consisting of a "detector" made of a piece of coke and a sewing-needle in an old medicine jar which he connected to an earphone. Hughes found that he could transmit the "click" of an electrical spark from his transmitter, first across his laboratory and then to other rooms in his house, using the metal fender of his living room firegrate as a primitive aerial. Finally, he set his clockwork-actuated transmitter going in the house, then walked up and down the street outside with his receiver, listening as the ticking noises that were being broadcast grew fainter and fainter until they died away at about 500 yards: a remarkable achievement with

such primitive low-powered apparatus. He was not fully aware of what he had done but, in fact, he had not only demonstrated practical "wire-less" broadcasting and reception – but had also built and used the world's first portable radio!

Hughes was certain that he had discovered a new phenomenon of "electrical conduction through the air" but hardly realized that he had produced tangible proof of the existence of aerial waves of a sort that scientists were only theorizing about. He was a member of the prestigious Royal Society, so he invited some celebrated scientist members and gave them a demonstration. They were initially impressed and astonished, but on reflection denied that he had done anything new. They declared that he had merely demonstrated already well-known "induction effects" which had nothing to do with what later became confirmed as real electro-

LEFT London-born Professor David Hughes, who made and successfully operated the world's first radio in 1879 when Marconi was only five years old. This discovery was unrecognized by the scientific establishment for almost two decades – until Marconi, just out of his teens, demonstrated it again.

The vital parts of Hughes' apparatus: an automatic transmitter made from coils, clock parts, and a battery (above) with a receiver consisting of a needle and piece of coke in a medicine bottle (left).

THE MARCONI STORY

Guglielmo Marconi, born of an Italian father and a Scottish-Irish mother in 1874, was in a way similar to David Hughes, in that he was a highly adventurous amateur scientist. He spent much of his early boyhood traveling with his family and his education was not of a specially academic kind. But he was excited by science from a very early age and soon became a keen experimenter. His parents turned the loft of their large villa near Bologna into a laboratory for him. There, in 1895, with the encouragement of his science teacher, he assembled electrical apparatus, set about doing his own experiments, and succeeded in demonstrating the transmitting and reception on radio for a distance of only a yard or two.

As has already been mentioned, Marconi did not "invent" radio, but assembled vital components developed by other researchers and experimenters into a working system according to principles which were already well known in scientific circles. Soon he was able to demonstrate a new, simple, thoroughly practical means of communication to important people who would be able to help him. He was lucky too in having

magnetic waves. He was deeply disappointed and decided not to publish anything about his experiments, but later wrote about how the professors had scoffed at the value of his work. Hughes was therefore denied his rightful "first" claim, and the development of radio was delayed by more than a decade, until the young Marconi burst upon the scene.

Driven by the kind of amateur zeal which dabbles in imaginative and freeranging experimentation rather than professionally structured academic research, Hughes achieved a breakthrough which his fellow-members of the Royal Society failed to appreciate.

a good imagination, a well-developed early business sense, and, above all, access to investment capital from a wealthy family who believed in him and had connections. Clever and ambitious, he was fired by a vision of being able to send messages through the air and was just twenty years old when he began experimenting with his transmitter and receiver, having recognized that all the elements of a system were already available: a high-voltage sparking-coil for producing "Hertzian" (wireless) waves, an aerial for transmitting them, and a crude and extremely simple receiver coupled to a buzzer or earphone borrowed from already established telephone apparatus.

Like other workers in the field at the time, Marconi could transmit only spark-modulated waves at that stage and he was still trying to improve on the short range of his transmitter. He set about improving his apparatus, adding primitive aerial arrangements, and found he could transmit across the room and then out of the window to his assistant's receiver in the garden below. Next, he made his equipment portable, took it into the open air and within a few months was able to transmit over more than a mile. Initially, only sparks could be transmitted. These sparks operated a bell or

could be heard in an earphone, but later a magnetic "printer" borrowed from the wired telegraph system was added, enabling the dots-and-dashes to be recorded on paper tape.

In less than a year, the 21-year-old Marconi had developed his system sufficiently to enable him to feel confident about demonstrating it and to devise ideas about its commercial exploitation. With a good deal of boldness he offered it to the Italian government, which, after inevitable bureaucratic delays, finally expressed no interest in it. Marconi was disappointed but decided to continue his endeavors. He saw a vital use for his system: the very serious need to provide communication between land and ships at sea, which were entirely out of touch when beyond the horizon. This lack of maritime contact had led to many disasters.

Not disheartened, Marconi secured the backing of his family and decided to travel to England accompanied by his mother, who had contacts there. He saw maritime communications as the main use for radio and recognized that, at that time, Britain was one of the world's great naval powers. He packed his equipment carefully but had to repair it when he arrived in London because suspicious customs officers, concerned about

THE HEART OF THE RADIO

The heart of any radio receiver is the "detector", which rectifies the tiny electrical signal coming down the aerial from the broadcasting transmitter.

RIGHT The evolution of detector devices is pictured here (clockwise): a lump of crystal and below it a catswhisker-and-crystal detector; the large valve with a later, smaller one: a solid-state crystal diode and a 3-legged transistor above it; a multi-combination chip. The old power-hungry valves were large and delicate, and ran hot. The devices that began to replace it after World War II used a minute amount of electricity, became smaller, were robust, and ran cold. Developing tremendously in complexity and sophistication, these solid-state devices sparked off the new Electronics Age with its amazing powers of "intelligence," affecting every aspect of contemporary life.

anarchist and anti-monarchist activities, thought it was a bomb, and ruined it while examining it. He then repaired it and set about successfully demonstrating it to the British Post Office, the Armed Forces, and other authorities.

The Press took a great deal of interest in Marconi, describing him as "the inventor of wireless" – a title he probably never claimed for himself but which was to become the firm belief of almost everyone.

Very soon, Marconi's equipment began to be installed in ships and military installations, but in those days radio was thought of as a very serious business.

Nobody could have guessed then that in just a few years it would become a universally accepted means of broadcasting to people in their homes all over the world.

Marconi patented his system in 1897. He set up a company to exploit it, and began to buy in other patents and the experts he needed to help him to develop his ideas. Then he set up the world's first radio factory – in England – as early as 1900. It produced a line of instruments and accessories for use by shipping, military, and commercial organizations, for there was as yet no concept of anything like a "home radio."

Although scientists were assembling their own small transmitters and receivers, these were only for experimental use and in any case were really simple two-way communicators. The possibility that anyone with a suitable receiver might be able to pick up a signal was at first seen to be a limitation, since radio was originally thought of as a personal one-to-one system with the advantage of privacy. This was not, of course, true: in fact one of the main advantages of radio broadcasting over the telegraph and telephone was that a signal from a single transmitter could be picked up by any listener within its range who had a suitable receiver. This was of particular importance in the case of SOS messages but was an obvious disadvantage in case the government, military, and commercial information authorities which required confidential communications. The early commercial exploiters of radio systems saw these groups as their potential customers, so they were not anxious to advertise this disadvantage. Conversely, their competitors were keen to emphasize it, so a public-relations battle ensued. Marconi came in for a lot of criticism from objectors who felt a monopoly would be contrary to the public interest, and accused him of making misleading and extravagant claims. One objector who set out to prove that Marconi's radio system was useless for confidential communications was Neville Maskelyne, son of the famous theatrical illusionist, who had invested in a rival company and waggishly put over his point in 1903, to the chagrin of Marconi and the amusement of the Press. He decided to play a wireless trick on Marconi's consultant, Dr. John Ambrose Fleming, who was giving a lecture at the prestigious Royal Institution in London. During the lecture a special message was to be received from the Marconi factory in Chelmsford, some 30 miles distant. Maskelyne set up a transmitter on the roof of his father's Egyptian Theatre close to the Royal Institution and interrupted their transmission with rude comments, including the limerick: "There was a young fellow from Italy, who diddled the public most prettily." The trick worked and the story got into all the newspapers.

Having secured the important technical patents and brought in the experts he needed, Marconi wanted sole control of radio communications between land and ships at sea, but this was overruled by an International Conference which decreed that radio communication

RIGHT A newspaper report of the world's first "wireless arrest" of 1910, when the notorious murderer Dr. Crippen was captured.

RIGHT The first air-to-ground radio being installed by Marconi's in a 1910 aircraft. A few years later, at the beginning of World War I, radio-equipped aircraft nicknamed "The Eyes in the Sky" were to be used as spotters for gun emplacements in the trenches.

must be available to all. A final blow to Marconi's bid for a monopoly of maritime communications came when the British government took over control of all Marconi's land stations in Britain. Soon, all transmitters in most countries came under licensing regulations – and it would not be long before licenses would be needed to operate receivers too.

THE FIRST HOME RADIO

Nobody knows exactly who first had a "home radio" installed, but it was probably no less than Queen Victoria, almost a century ago! A grand old lady of 77, she was then celebrating the Diamond Jubilee of her reign, had always taken interest in new ideas, and in 1897 already had a telephone – installed and demonstrated to her by the inventor, Alexander Graham Bell. In the following year she commissioned Marconi to install "wireless" apparatus at Osborne House on the Isle of Wight off the south coast of England. She used it to receive bulletins on the progress of her injured son the Prince of Wales (later Edward VII) while he was convalescing aboard the radio-equipped Royal Yacht which was moored a few miles away in Cowes harbor.

The Press pounced on this as a good story, dubbing Marconi as "the inventor of radio." The fact that he got into the public eye, developed a flair for self-publicity, and engendered so much excitement for his enterprise throughout the world, put him ahead of his competitors. Marconi continued to increase the range of his equipment and pulled off another publicity coup at the end of 1901 by transmitting across the Atlantic – a feat that demonstrated to the world at large that the Radio Age had really begun.

He got into the papers again in 1910 when the world read of the first "arrest by wireless" at sea. By that time, Marconi had installed radios in a great many ships. On one of them, the liner SS *Montrose*, the infamous murderer Dr. Crippen was escaping with his mistress Ethel le Neve. The captain recognized them and sent a message back to London to arrange for their arrest when the ship sailed into port.

Meanwhile radio had caught people's imagination and was being developed along less narrow lines: there were experiments in several parts of the world to foster radio as a means of disseminating information to the general public, and there were even some early attempts at broadcasting entertainment.

LEFT The "Wireless Den" of London amateur transmitting enthusiast F. Livingstone in 1923. Nicknamed "Radio Hams," these experimenters played a great part in developing equipment, popularizing radios and fighting governments for public right of access to the airwaves.

The outbreak of World War I in 1914 gave further impetus to the development of radio. It was already installed in naval ships but was soon being used in balloons, airships, and aeroplanes nicknamed "The Spies in the Sky" which acted as "spotters" for gunners in the trenches. At the end of the war, a large number of engineers with experience of radio were released from the forces and they played an important part in the rapid development of radio in many different directions in the Twenties.

Few of these skilled ex-servicemen could expect jobs in the infant radio industry, but having been fired with enthusiasm they continued radio as a hobby. They made their own receivers and in some cases even constructed and operated their own transmitters at home – utilizing the large amount of ex-government surplus equipment which appeared very cheaply on the market. These early radio amateurs (nicknamed "hams") not only helped to improve and develop equipment but also encouraged interest among the general population for broadcasting, and they defended people's rights to the airwaves. Governments and other authorities which already controlled post and telegraph services, soon began to view the radio amateurs with a certain amount of alarm and suspicion, regarding them as potentially anarchic and subversive agencies which

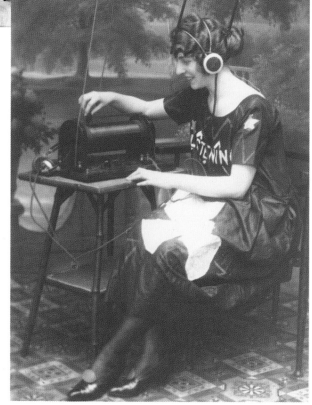

ABOVE This Twenties flapper made her own crystal set from a cardboard tube, wire and a catswhisker-and-crystal. Wearing earphones and an aerial on her head, she proudly posed in a London photographer's studio against an oil-painting background for this 1922 picture postcard.

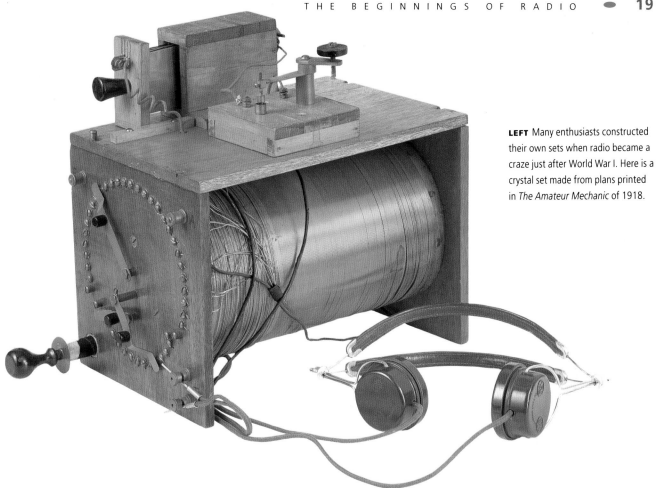

LEFT Many enthusiasts constructed their own sets when radio became a craze just after World War I. Here is a crystal set made from plans printed in *The Amateur Mechanic* of 1918.

needed strict controls. They argued that uncontrolled broadcasting might crowd, jam, or otherwise interfere with official communications. This argument had some validity in those days, so the amateurs were given "license" rather than freedom to broadcast.

But while governments and officialdom regarded radio as a serious service and quickly took control of it, the general public saw its possibilities as a home-entertainment medium. People were caught up in the rush to possess for themselves the magical toy which could miraculously tune in to music and voices in faraway places. Music halls, bars, and movie theaters feared for their own survival. Even the piano in the parlor began to lose its traditional status as a focus for family entertainment.

Well-to-do people were able to buy commercially produced receivers, but these were very expensive, costing at least three times an average weekly wage, so the less privileged made their own sets at home, many taking up evening classes at "Working Men's Institutes" to learn the necessary skills, which they hoped in those

Depression years might lead to employment in the new and fast-developing industry. Luckily for them, the early radio receivers were simple to construct – especially the first "crystal sets," consisting of three or four working parts and a pair of earphones, putting them within the reach of even a schoolboy's pocket-money.

The sets needed patience to tune in and, even with a very long, elevated aerial, were unreliable, but the cult of making a good set and becoming an expert at tuning-in was all part of the fun and one-upmanship. So keen were the early experimenters that they erected 100-foot aerials in the back yards of their houses, strung from the eaves to poles like washing-lines, in the hope of hearing stations far away.

In the very early days there weren't even any proper broadcasts to be tuned-in, but simply the pips of Morse code from ships at sea, test signals from experimental transmitters, or perhaps time-signals sent out from high buildings like the Eiffel Tower to enable people to set their clocks. Yet enthusiasts would sit up all night for the thrill of a single distant contact.

RIGHT The Goltone Super crystal set of 1925 – probably the first ever bakelite radio.

CRYSTAL SETS - AND HOW THEY WORKED

Many of these primitive receivers were crystal sets, consisting of three or four components screwed to an old wooden bread-board. The most important component was a "catswhisker" of coiled wire and a piece of Galena crystal (easily found in the slag-heaps of lead mines) which could detect or rectify the current from a wireless wave coming down from an aerial (antenna) wire. The only other components needed were a simple tuning device and an earphone which converted the signal into sound waves. No batteries were needed to activate it because all the power required – a very tiny amount – came down the aerial from the broadcasting station.

For most people, home-construction was a necessity, since factory-made sets were so expensive. These commercial sets still had only a few components, but were cased in fancy french-polished boxes made from exotic woods and came with impressive diagrams and instructions. They also came with spare crystals packed in cotton-wool in fancy tins and with names like "The Mighty Atom" and "The Neutron," together with "catswhiskers" of gold and silver. The crystals were common minerals and the amount of precious metals microscopic, but the presentation added to the pseudo-scientific impression of it all. This was an early form of clever wireless marketing which was to be employed by manufacturers of domestic radios as well.

For the price of about half a week's wages, the average man could add a tube to his crystal set to increase its output power and range, and let him hear the messages passing between ships at sea and perhaps the broadcasts being sent on low power by amateur transmitters.

It was the arrival of the vacuum tube which provided the means for real progress in the technology of radio. This device, which enabled more complex development of electronic circuits, better detection and tuning, and above all increasing amplification of signals and better quality, was developed like radio itself, in a process of leaps and bounds by workers in various countries, rather than by a single person.

The American inventor Thomas Edison – the man who built and demonstrated the first practical phonograph – started off the evolutionary process of the tube when he observed a strange effect while experimenting with his electric light bulb in 1883. He found that when he installed an extra electrode – a flat metal plate – into the bulb near the glowing filament, the device acted as a kind of one-way switch for electric currents. The action was analogous to the one-way tube in a water-system, and this "Edison Effect" later enabled it to be developed in a way which made the old detection devices such as the "crystal-with-catswhisker" entirely obsolete, and revolutionized radio. In his "Invention Factory," the brilliant and enterprising Edison was busy on so many other projects that he put the idea aside, although he went on perfecting his light bulb, unaware of its huge significance. It was fortunate that he demonstrated the device to visitors to his laboratory, in particular to the British physicist John Ambrose Fleming, who was later to work with Marconi on his transatlantic radio project. It was not, however, until some twenty years after he had first seen it that

Fleming himself began experimenting with electric light bulbs with various extra electrodes, leading to the development of his own "oscillation valve" which he recognized would work as a detector and accurate responder to radio signals. It was improved in 1906 by the American Dr. Lee De Forest, whose "audion valve" could be controlled by use of an extra metal "grid" by which its behavior could be controlled, and which gave it the important property of being able to amplify weak signals. Without the development of the tube, we might speculate that we would still have only the crystal set, with its earphone reception, as our radio, although strangely enough, after half a century of supremacy, the tube itself became obsolete when the transistor arrived. Stranger still to relate, the transistor, which revolutionized almost everything in the world of radio in 1951, is actually a closer relation of the old "catswhisker-and-crystal" than the superceded tube. This occurrence again proves the old adage that history rarely repeats itself!

AESTHETIC DEVELOPMENT

The aesthetics of radio developed alongside its technology and around 1930 there was a sudden leap as the battery wireless set, with its long aerial and separate loudspeaker, gave way to the electrically driven, self-contained radio. By the early Thirties it was totally practical, cheap to buy, simple to operate, thoroughly domesticated – and even portable enough to be installed in automobiles and taken on picnics.

In the Thirties, radio ended its infancy as a scientific toy and became regarded as a necessity for serious communication work and as a means of entertainment for the home.

By the beginning of the Thirties almost all the important developments in the technology of receiver design had taken place (with the exception of the transistor and the printed circuit, which did not appear until after World War II). This resulted in a basic standard group of types of receivers. These tried-and-

THE EVOLUTION OF THE RADIO

The increase in complexity of the modern radio has led to a decrease in size, seen clearly here in the evolution from the original, cumbersome apparatus of the Home Radio of the Twenties to the many miniaturized versions in use today (overleaf).

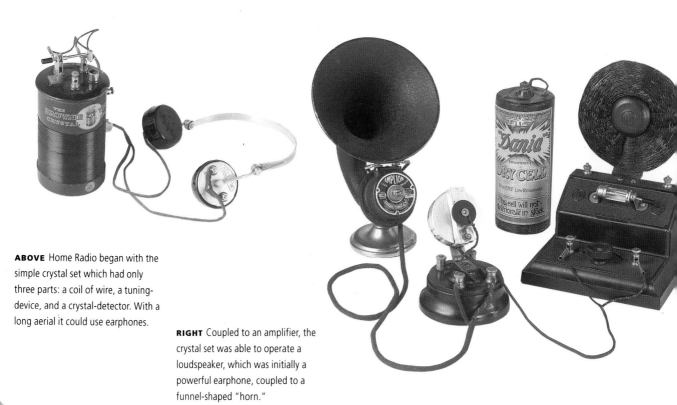

ABOVE Home Radio began with the simple crystal set which had only three parts: a coil of wire, a tuning-device, and a crystal-detector. With a long aerial it could use earphones.

RIGHT Coupled to an amplifier, the crystal set was able to operate a loudspeaker, which was initially a powerful earphone, coupled to a funnel-shaped "horn."

tested receiver designs were adopted by the proliferating small radio manufacturers, but to survive and compete with each other they were forced either to offer extra features such as superior cabinets to attract buyers, or to cut costs and bring down their prices. For the smaller manufacturers, providing extra features was possible only by specialization in luxury sets and low production, with expensive price-tag models where the available market segment was small. There was only room for a few such contestants. Conversely, these small firms with limited production capacity were unable to compete with the medium-sized firms in the price war for the larger, general section of the market. Caught between these difficult alternatives, many of

them went out of business, abandoned radio production, or were absorbed by the fast-growing big players in the game as the radio boom really got going.

After radios began to be mass-produced, the need arose to increase production continually, and this was done by marketing a large variety of new models to encourage customers to update their radios and to purchase more than one receiver. Radios were produced for bedrooms, kitchens, and automobiles, as well as for the living room. The technical features developed too – tuning could be automatic instead of manual, the number of stations receivable increased, and there were improvements in sound quality as well as the aesthetics of cabinet design.

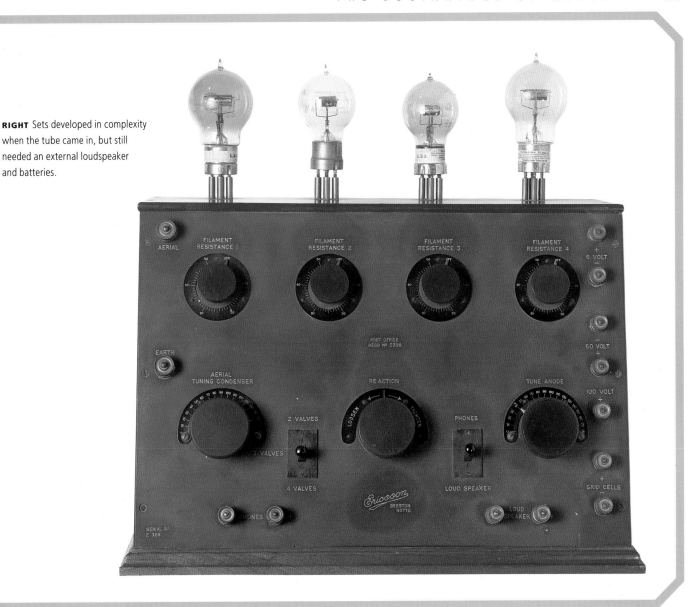

RIGHT Sets developed in complexity when the tube came in, but still needed an external loudspeaker and batteries.

All these features formed part of the Golden Age of Radio, the crowning glory of which is thought of by many people to be the artistic creation of cabinets in bakelite and its derivatives. The first period of three decades was ended by the outbreak of World War II. It was then that the radio was forced to give up the jazzy, giddy fantasy of its Depression-born escapism, exchanging its bright colors for a uniform green and gray to join the forces and changing its tune from foxtrot to Morse code and the occasional patriotic song. The radio, on active service, became deadly serious again and dressed the part as communicator, morale-booster, and propaganda-machine in strictly functional khaki metal boxes.

After the war, revivalist radios in prewar ornamental wooden boxes and in bakelite began to reappear, but it was an echo of a dying sigh: the Golden Age of Radio had really gone for ever. What lay ahead was a sort of anti-art in radio that displayed a fear of decoration and adopted a fashion which became more and more minimalist, until the cabinet disappeared altogether inside matt-black marketing packages designed to give a hi-tech image.

Another more forceful reason for the supplanting of the decorative radio as the focal point of the living room was the dramatic arrival of the television set after the war. The radio had achieved the status of a decorative object in the home, welcomed as much for

RIGHT At the end of the Twenties, sets became self-contained with all the parts in one box and worked from main electricity.

its own beauty as for the service it gave as a purveyor of news, music, education, and entertainment. It was "a friend in the corner" – and one to be seen as well as heard. By contrast, the television was a plain box, welcomed for the pictures it presented rather than as a decorative object. The moving pictures it showed proved to be far more arresting than the sound pictures that came from the radio, and demanded the full attention of the viewer in contrast to the radio which let people do other things while they listened. It was exciting and new, and cheaper than going out to the movies. The radio became a mere household machine like the washing-machine and the vacuum-cleaner in a boringly functional garb. The radical changes in

technology that came with the invention of the transistor, the microchip, the printed circuit, and extreme miniaturization ousted the large, power-hungry, hot-valve "works" of the old technology. The radio soon got incorporated into stacked music systems in the home and entirely lost its original character and separate identity.

Soon, people were prophesying that television would rapidly kill off radio, newspapers, magazines, books, records, schools, and gramophone records. But television's very popularity generated its own critics: philosopher Bertrand Russell dismissed it with the words, "It will be of no importance in your lifetime and mine," and architect Frank Lloyd Wright described it as

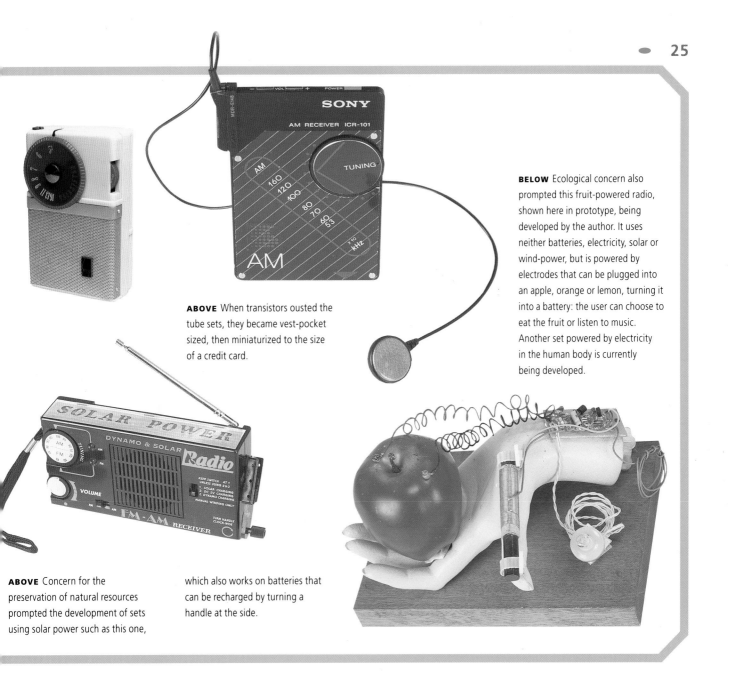

ABOVE When transistors ousted the tube sets, they became vest-pocket sized, then miniaturized to the size of a credit card.

BELOW Ecological concern also prompted this fruit-powered radio, shown here in prototype, being developed by the author. It uses neither batteries, electricity, solar or wind-power, but is powered by electrodes that can be plugged into an apple, orange or lemon, turning it into a battery: the user can choose to eat the fruit or listen to music. Another set powered by electricity in the human body is currently being developed.

ABOVE Concern for the preservation of natural resources prompted the development of sets using solar power such as this one,

which also works on batteries that can be recharged by turning a handle at the side.

"chewing-gum for your eyes." But as was the case with radio, television stimulated interest in other media rather than threatened them. Other media were recognized as having the very special quality of allowing the participant to use his imagination – many people still say "We prefer radio: the pictures are better!"

Radio has also proved to be adaptable to quite new specializations which make it a more portable and personal medium than television. Personalization of the technology has meant a democratization of radio, producing transmitter-and-receiver devices which are no longer under the exclusive control of government and commercial national and international broadcasting authorities, but which give ordinary individuals access to

the airwaves. At the same time as providing this new freedom, the continued progress in the use of radiowaves has also brought mixed blessings – one example being the disturbing extension of radio surveillance, which on the one hand promises tighter control of criminal activities and on the other warns that Big Brother is already watching us all.

Today's electronic wizardry, for good or ill, all stems from the beginning of the Radio Age. The radio has long since moved outside of our homes, where we can to some extent choose what we wish to experience, and into the street, where it can become intrusive – in the form of "ghetto blasters" and the powerful music systems of motorists.

CHAPTER TWO

A Short History of the
Radio Cabinet

efore Marconi established the world's first radio factory in England in 1900, experimenters requiring equipment would commission items to be made for them by well-established small manufacturers of general scientific equipment and makers of electric apparatus who were already producing the standard items then needed for radio work. Most of the separate pieces required for wireless were already in existence before the turn of the century because they were standard equipment in scientific laboratories – items like induction-coils, batteries, spark-gaps, buzzers, earphones, and detector devices. To make a transmitter or a receiver, only a few components were required and these could be wired together on the laboratory table. At that time wireless equipment consisted of this loose assembly of separate parts and was not installed in a cabinet of any kind as a complete unit that could be thought of as a radio.

Soon, the Marconi Company and its competitors were supplying complete kits and they began to install some of the discrete parts in a series of boxes to simplify interconnections and to enable more robust, portable, and easily installed outfits to be supplied to customers. Thus, the original "wireless set" of parts became the boxed form first called the "wireless," then the "radio."

At first, these early radio cabinets were strictly functional; they were merely boxes to house the "works" of radio. There was no sense of "design" about them, but in the tradition of nineteenth-century scientific instruments they were made by hand by craftsmen, using materials of high quality. In the days before sophisticated automatic factory machinery and production-line manufacture, the designer – specially in small firms – was also the craftsman who actually constructed the object.

SMALL FIRMS

Such small factories in the embryo radio-cabinet and components industry were engaged in low-volume, high-quality work. They employed skilled woodworkers, metal-workers, finishers, electrical and mechanical engineers, and even decorative specialists whose general skills enabled them to adapt easily to radio production.

More and more small engineering, electrical, woodworking, and metalworking firms were following Marconi's lead and beginning to change over their production to radio components. Standardization of components enabled them to turn out lines of models from simple crystal sets to multi-tube ones as well as units like tuners, amplifiers, and power supplies with separate loudspeakers, but until about 1930 there was no concept of an all-in-one radio.

Despite this rapidly expanding manufacturing activity in the first two decades of the twentieth century, the production of radio equipment remained on a small scale, being generally confined to "serious" uses such as saving lives at sea and for military and commercial applications. In the early days, there was little thought of exploiting radio for less serious purposes – even in the mind of Marconi himself. Indeed, the early pioneers would have considered any suggestion that it might be used merely to entertain the masses as an irreverent devaluation of their concept of radio as a saver of lives, an instrument of government, a powerful weapon of war, and a vital tool of commerce.

The early forms of apparatus soon became obsolete as the technology became more and more complex. The simple crystal set, which had a limited range and would only work an earphone, was

ABOVE Staff parading for the photographer inside the world's first radio factory, set up by Marconi in Chelmsford, England, in 1900.

superseded when the real development of radio began as a result of the development of the "thermionic valve." This powerful detection and amplification device, looking rather like an electric light bulb, provided amplification for long-distance and loud-speaker reception. It led to a complex and sophisticated development of the whole "works" of the radio in the Thirties, leading to the introduction of a perfected and more-or-less standard design that lasted until just after World War II when there was another revolution which gave birth to modern electronics.

THE DOMESTICATION OF RADIO

Many workers in smaller industries like the burgeoning radio factories had not yet been totally de-skilled by the production-line techniques of the Twenties. They still possessed crafts skills that were welcomed in the radio workshop, enabling them to construct and assemble the "works" of radios as well as the cabinets. They were

ABOVE An early valve.

RIGHT Victorian values endured into the next century. "Traditional quality and modern craftsmanship are happily combined" in the unusual radio advertised here, manufactured in small quantities by Remler of Los Angeles in 1936. (Radio pictured on page 81.)

"STERLING" REMLER QUALITY SINCE 1918

In years gone by Remler "Builder of world's first superheterodyne" has striven unremittingly to earn for its products the approval of those who appreciate quality merchandise. This undeviating adherence to the highest ideals of manufacture and design has resulted in a series of radio receivers of literally superlative worth.

The Remler Scottie is no exception. It is the climax of months of painstaking research and skilled engineering. In it traditional quality and modern craftsmanship are happily combined. It is in every way worthy of Remler—the name that personifies quality in radio.

Remler Scottie sold and sponsored by

Made by Remler Co., Ltd., 2101 Bryant St., San Francisco

Companionable as a Puppy—the

REMLER SCOTTIE

RIGHT A 1935 bakelite radio imitating the look of wood. In the tradition of the old craftsmen, the new designers strived to make the cabinet an acceptable sight in the parlor.

practical people who designed what they made at the workbench and were still allowed an aesthetic input, since separate design departments employing drawing-board designers existed only in large factories. Their philosophy was that of the Arts and Crafts movement, which in the words of William Morris advocated: "Have nothing in your homes which you do not know to be useful and believe to be beautiful."

This old-fashioned system naturally produced radio cabinets that were of traditional style rather than more innovative designs that belonged to a dawning "Machine Age." The old craftsmen knew how to deal with such innovations as the automobile because they could base their designs on the horse-drawn carriage, but when suddenly faced with the radio they must have been perplexed. It was something entirely new – the first electronic entertainment machine to enter the home – and nobody had much idea of what its appearance should be. The radio at that time was simply a functional box-of-tricks and that's how it looked – like the scientific instrument it really was.

The hallowed parlor, with its atmosphere halfway between a church and a museum, was immediately judged to be the wrong place for the scientific-looking embryo radio, with its knobs, dials, glowing valves, glistening crystals, coils, and wires. Yet people really wanted what the magic radio would bring and faced a dilemma: how could it be domesticated?

Well, they did what the Victorians always did with problems raised by the vital social necessity of keeping up appearances – they either ignored what they found unpalatable or carefully hid it away. The useful but unmentionable chamber-pot was hidden in an ornate cupboard or decorated as nicely as a Wedgwood vase; the well-patronized upmarket brothel was disguised as an overdressed ladies' "At Home;" and the essential public lavatory pretended to be a Roman temple.

It was true that other machines had already entered the home, like the hand-pumped vacuum-cleaner and the hand-operated washing-machine, the sewing-machine, the rotary knife-sharpener, the meat-grinder, and the bacon-slicer. But all these machines were installed "below-stairs" for the servants' use, being hardly appropriate for the living room, so apart from a little gilt decoration they retained the look of machines.

The only domestic machines allowed in the parlor were such items as the pianola, the harmonium, the music-box, the wind-up phonograph and gramophone (which were also to be electrified), and the first piece of electric technology to enter the home: the telephone. These machines had already been made acceptable by embellishment or disguise as pieces of furniture.

Subtle marketing embellishments soon made these things acceptable in the parlor; classical gold-leaf decorations, brass knobs, impressive name-plates, fancy floral transfers, and exotic wood veneers. Designs were borrowed from several periods, from the simple and

understated elegance of Chippendale to the ugly excesses of Victorian Gothic.

But when the radio suddenly appeared, it was a special challenge for the designers, who could hardly have guessed that it would soon stand alone as the undisguised focal point in almost every home. It was also an extremely expensive device, locked up like a tea-caddy or a tantalus, to be operated only by the master-of-the-house. He kept the key on his watch-chain and considered himself the only person clever enough to work it; a distinction not then accorded even to the lady-of-the-house (who was much too ladylike for such an occupation), or the servants who were presumed too ignorant. The butler, however, who had already been allocated the task of winding-up the gramophone, might be summoned to tune the radio under instruction.

In embryo form, the early radio consisted of a wooden board on which were mounted a few components like coils, glass bulbs, rotary-vane condensers, dials, and tuning-knobs. It was all connected together by a spider's-web of wires and attached to a pack of large batteries, a pair of earphones or a loudspeaker, but requiring even more wire to provide an aerial and ground, this ugly assembly seemed chaotic, but made sense electronically, so the components could not be easily changed or repositioned simply to please the eye of the cosmeticist.

The designer's "decorate, disguise or conceal" formula also survived into the manufacture of bakelite radios.

LEFT General Electric's Jewel Box, a good example of bakelite imitating a well-crafted wooden cabinet, concealing the radio when closed.

BELOW The 1934 Colonial, a bakelite radio masquerading as a globe.

There were, however, some precedents in the process of mechanical domestication. The telephone, with its strange trumpet and generator-handle, and the phonograph and gramophone, with its spinning wheels, winder, and tin horn, were even more like machines than the new radio apparatus, yet they were tamed by the use of the designer's "decorate, disguise or conceal" formula.

Decoration came first, usually consisting of the use of attractive polished metals and applied fancy embellishment in classical style, making the objects little works of art in themselves to obscure their practical functionality – much in the same way as a common workhorse traction-engine can be magically transformed into a colorful fairground attraction with brightly colored paint and flickering lights. Next, the designers turned their attention to the art of concealment, putting the mechanism into well-crafted cabinets made from expensive and exotic woods. This made them objects of beauty, so concealing their real function and making telephones look like ornamental clocks and gramophones like miniature grand-pianos.

With the advent of the loudspeaker, speaker apertures were covered with decorative silk and framed with fretworked ornamental grilles, seen here in two radios made in Great Britain in 1931.

THE EVOLUTION OF THE RADIO CABINET

These techniques were gradually used to domesticate the radio which was still a mess of separate parts and yards of wire, even by the mid-Twenties. Most of the components could be put into a box and, by then, a loudspeaker had been devised. This was achieved simply by adding a gramophone-horn to the earphone which had been used for personal listening. Later, flat "cone" loudspeakers in square wooden boxes replaced decorated freestanding horns. To get enough volume to drive increasingly larger loudspeakers, an amplifier was added. This used power-hungry tubes to make sounds louder – and these in turn demanded large, heavy and expensive batteries to run, since few people then had domestic electric power. Radios were therefore beginning to become quite large and more complicated, so that they could not be easily disguised or hidden away. As a result, the designers made them part of the

acceptable furniture of the living room, constructing them as cupboards, chests on legs, dressers, writing-boxes, and desks. This furniture-style treatment continued for some years, so that it was a long time before the radio became a separate object in its own right in the home.

This change happened quite abruptly, when a craftsman had the idea of putting all the electronic components of the radio, including the batteries and an aerial wound on a frame into a square wooden box, then cutting a round hole in the front for the loudspeaker, creating the first entirely self-contained cabinet radio. Next, the speaker aperture was covered with decorative silk cloth and framed with fretwork ornamental grilles of fancy woods and metals. At first, these were in Art Nouveau patterns of interwoven natural forms and later of designs derived from recently discovered Egyptian tomb and Mayan temple art forms which pointed the way to an Art Deco style.

The real radio cabinet had arrived! At more or less the same time, there began a rapid and profound development of the technology of the receiver itself.

THE GOLDEN AGE OF THE CABINET

The use of bakelite freed the designer's hand and led to a Golden Age in cabinet design.
This came to an end when radios became pocket-sized and needed only cases – or began
to be hidden inside stacking music centres.

BELOW Cabinets for domestic sets
from the beginning of the Twenties
were made of fine woods, like this
crystal set, for earphone use.

ABOVE Then came the tube set, still
in a wooden cabinet and needing an
external loudspeaker.

RIGHT When the "works" were completed with a flat "cone" loudspeaker, everything went into a single cabinet. Cabinets began to develop as an art form in wood, like this cubist-style radio of the Thirties.

BELOW The use of bakelite and the many plastics that followed it freed the designer to make cabinets in a true variety of forms, like this Australian Healing Moderne of 1950.

ABOVE After World War II, for a brief period radios were designed as decorative objects – some were incredibly kitsch, like this spaceship design made in Great Britain in 1947.

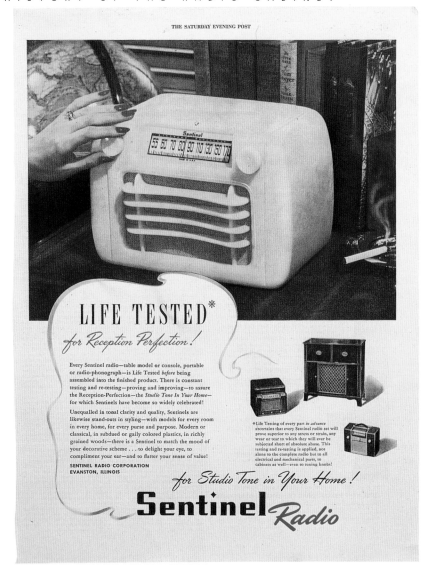

RIGHT An advertisement for a 1945 Sentinel radio, geared to appeal to the customer's "sense of value." Just as importantly, it tells us, there is a Sentinel to match the decorative scheme of any room!

INDUSTRIAL DEVELOPMENT AND MASS PRODUCTION

Until the boom came, radios had been produced only for the rich and the comparatively well-to-do middle-classes while the less affluent either had to construct their own sets or go without. Now, everybody wanted to own a radio and phenomenally increasing demand, coupled with the necessity of reducing prices to make good radios within the reach of all, forced manufacturers to adopt mass production techniques to produce the "chassis" (or "works") of the radios. These high-volume techniques required increasingly sophisticated mechanization and the conversion of the labor force from a large army of multi-skilled autonomous workers into a rapidly reducing group of virtually unskilled machine-feeders. Increased production was met by increasing mechanization which in turn demanded standardization of components and division of labor into separate groups which produced only discrete sections of the final product. These were brought together on specialist assembly-lines. The new manufacturing process managed to increase profits for the manufacturer and to decrease prices for the consumer, which in turn stimulated further increases in production. Such a system of production had been identified in the mid-nineteenth century by Karl Marx in *Das Kapital*, in which he described how "machines had been made to do the work of men" and how the workshop was "an engine, the parts of which were men." Earlier, in England, Josiah Wedgwood, the Staffordshire pottery entrepreneur, had advocated "making machines of men."

These major developments required huge capital investment which was available only to big manufacturing concerns and, as these developed, the largely unmechanized smaller firms producing individual craftsman-style radios became uneconomic and uncompetitive, and were either taken over or forced out of business.

Profound changes in production techniques began to force changes not only in the technology of the radio but also the design of the cabinet. This factor is often overlooked by design-orientated historians who overemphasize the artistic input of "famous" designers in changing design, while disregarding the economic and cultural determinants of change.

Certainly creative and imaginative designers produced innovative work influenced by contemporary movements in painting and sculpture; but the impetus for change actually came from the imposed demands of factory economics. Selling the new designs required aggressive promotion and modern marketing techniques just as much as a sensitivity to market demand: sometimes a product was tailored to suit an idea of

what the customer wanted but more often the product came first and the customer was persuaded to like it. There were also examples of outlandish and unsaleable radios being produced despite prior market research. One apocryphal story about such marketing is still told by veterans of the early days of the radio industry. It tells of a case in which potential customers were asked what sort of radio they wanted and replied that it should be "elegant, efficient and economic," but when it was produced they declined to buy it. When the same customers were asked what they thought "the masses" would buy, they replied that it would be something "big, cheap and flashy, got up to look expensive," but when it was produced, they bought it.

But there was a boom time in the Thirties which lasted until the majority of households were equipped and an industry which had overproduced realized it was necessary to find ways to keep profits rolling in. Many ideas were used to tackle the problem. One device was an advertising scheme to encourage people to have "A radio in every room," with the offer of receivers in colors to match kitchens and bedrooms as well as "Mickey Mouse" designs for children's rooms. Other promotions included portable and automobile radios. Less ethical market-boosting devices were said also to have been used, such as "built-in-obsolescence"

LEFT A Mickey Mouse design for a children's room – an American Emerson radio molded in "repwood" (plastic!).

CABINET MATERIALS

Radios have been made from many different materials in their long history. Here are some of the most unusual ones.

ABOVE Cardboard A 1922 crystal set concealed in a picture-postcard, produced in England and Germany, which actually works an earphone.

RIGHT China This is probably the world's most rare crystal set, made by the Scottish Britannia works in 1923 in the form of a Persian boy reading the *Ruba'iyat of Omar Khayyam*. The controls are at the back.

resulting from rapid changes in style to insure that designs would soon go out of fashion; and the use of poor quality components which rapidly failed.

Such people as the craftsman-designer, the skilled assembler, and the electronic engineer were doomed in an age in which productivity was king and individuality would no longer be valued except in the elite echelons of the fine arts. The handmade creations inspired by the Arts and Crafts movement were soon to be replaced by synthetic manufactured objects in which "art" metamorphosed into lines of regimented yet ever-changing "styles," apparently market-led but actually leading the market by the nose as a result of clever promotion.

The mass production technique involved designing

a line of standard models of the "works" (chassis) of the radio, using interchangeable parts which machines could turn out rapidly and economically. Cabinets, however, were a different matter. Until the early Thirties, they were made mostly of wood. Though attractive, wood by its very nature was unsuitable for mass production. Designers were restricted to mainly square-box shapes, with added decoration to break the monotonous straight lines, because wood does not lend itself to bending and is expensive and wasteful to carve into curved shapes. Wood also distorts with changes in temperature and humidity, which makes it impossible to produce cabinets to the same degree of accuracy as the metal chassis to be fitted into them, leading to assembly

LEFT Celluloid A 1922 crystal set in the form of a book in imitation tortoiseshell.

BELOW Synthetics Two examples of radios made by British Ultra in 1953. The imitation leather cases were nicknamed "Fake Snake" and "Mock Croc."

LEFT Perspex A British Pye "Personal Portable" radio of 1946. A persistently quoted story about the set says that it was hastily withdrawn from sale and stocks were burned at the factory because the Art Deco "sunrise" design was in bad taste, being similar to the wartime flag of Japan – Britain's enemy of a year before. The story may be apocryphal: dealers disliked the set because it was badly made from cheap components, unreliable, and almost impossible to repair, so the story may have been a smokescreen to hide a designer gaffe.

problems. Construction in wood also involves a large number of separate operations, and the wood subsequently needs to be finished with stains and polishes.

THE NEED FOR PLASTICS

Early radio manufacturers soon came up against these problems, and must have dreamed of finding a material that would replace wood, yet would still be seen as beautiful – particularly in domestic surroundings. Maybe they yearned for a kind of wooden "pastry dough" that could be cut and molded into any desired form and then be "cooked " to a permanent, ready-finished state. In fact, there were a few experiments at making small objects in "plastic wood" – a mixture of glue and wood-flour. Only a very few radio cabinets were ever made of this because it presented molding problems, and it was not received with enthusiasm since the material looked like compressed cork and bore none of the graining and color which makes wood so attractive.

It was the appearance of plastics with special characteristics which gave freedom to the designer, for he could now shape his material to suit his own purposes rather than having to put up with the limitations of "natural" materials. The result was that he could begin with his ideal aesthetic, produce a practical and economic design, and then tailor his plastic material to suit: the material could be hard or

CABINET MATERIALS

RIGHT Cloth A 1957 "Handbag"
radio by the British Jewel Company,
in a fiber and cloth case.

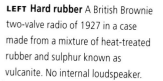

LEFT Hard rubber A British Brownie
two-valve radio of 1927 in a case
made from a mixture of heat-treated
rubber and sulphur known as
vulcanite. No internal loudspeaker.

soft, rigid or flexible, transparent, translucent or opaque, heat-resistant, and in any desired color or finish. It could be formed into shapes in the way that a pastry cook stamps out designs with simple tools, or poured into molds and turned out like fruit jellies. It could imitate other materials – even "natural" ones like wood, ivory, gemstones, tortoiseshell, animal skins, and metals. It was comparably cheap and ideally suited to mass production techniques. Radio cabinets could be turned out rapidly and already complete, thereby tying in well with the assembly-line process.

Like the advocates of "Arts and Crafts" before them, some designers – and popular opinion too – complained that mechanization would take the heart and soul out of design, killing off the last vestiges of craftsmanship. At the same time the more enterprising avant-garde designers welcomed the new materials as a liberating force. They saw plastics as a way of changing

design aesthetics to serve a mass market with innovative products. Plastics also gave them an opportunity to deliver "style" to increasingly culturally literate ordinary people who were beginning to be excited by the promise of a dawning Brave New World of technology which would soon emancipate them from manual labor, and give them cut-price culture and the leisure time to indulge themselves in it.

Initially, only fellow initiates to "Modernism" shared the enthusiasm of these pioneer architects and designers of consumer goods. But soon their products began to become established as a distinct and marketable "style" and their rich and sophisticated patrons embraced this as a social indicator to demonstrate their elite, cultural superiority.

Conversely, the general masses at first rejected such modernity despite its hedonistic and democratic promise, dismissing it with the same contempt and

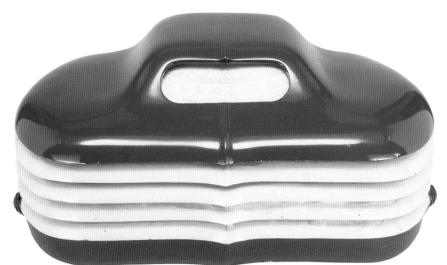

LEFT Wood The "Iberia" post-war valve radio, made in wood and paint sprayed, pretending to be plastic.

LEFT Metal A "Shoulder-Bag" radio by British Romac in a metal case, with its aerial concealed in the strap.

RIGHT Glass Tube radio constructed in a real glass bottle with bakelite plinth, "bottled" by Radio Development and Research of America, giving "mellow music to soothe your spirits and 100 proof performance."

ridicule with which they confronted "Modern Art." Just as they received avant-garde painters like Picasso with amusement and derision, they eschewed the works of designers such as those of the innovative Bauhaus, dismissing them as suppliers of props for pretentious people posing in fashionable studios rather than living in real homes.

Wood, the traditional material for furniture, remained popular for cabinets even into the Sixties because of its beautiful appearance, its sense of "naturalness," its feeling of "warmth" and cosiness, and the fact that it seemed entirely at home with the popular furniture styles of the majority of living rooms. Plastics continued to try to imitate wood and, at the same time, there was a brief attempt to use wood to make cabinets in the styles which plastics designers were using. This was done by employing new techniques of "steam-forming" plywood into curved forms and building up stepped and rounded sections from carved shapes.

But it soon became evident that the "wood versus plastics" battle was over and that plastics had won. Not long afterward World War II stopped the development of the radio, which went into uniform and became starkly functional in appearance. When the war was over, old ideas about cabinet design were resurrected and new ones developed for a time, but by then changes in the whole technology of radio came with the arrival of the transistor, the printed circuit, and the microchip which miniaturized the "works" of the set in such a dramatic way that elaborate cabinets were no longer needed. The Golden Age of the radio was gone for ever: except for today's booming "nostalgia" market of reproduction old-style cabinets with up-to-date electronics inside them – regarded by many people as hideous anachronisms.

CHAPTER THREE

A Short History of Bakelite and other plastics

Addison

We now live in a "Plastics Age," which is popularly thought of as being almost synonymous with the "Machine Age" within the "Age of Modernity" – but the story of plastics and its technology goes back a very long way indeed.

Animal, vegetable and mineral resources have all been exploited to provide plastics. "Natural" plastics like amber, horn, and tortoiseshell were ready-to-use. Others, like hard rubber and bitumen are semi-synthetics: chemically modified natural substances, but the vast majority of plastics used today – totally synthetic products of the twentieth-century laboratory – are based on oil and natural gas. All are distinguished by their ability to be shaped by various processes and by their particular molecular structure. Some are called "thermoplastics" because they can be softened and reshaped by heat and pressure processes; others are known as "thermosetting" because they remain permanently "set" when formed.

Modern plastics have followed from achievements in chemical laboratories in the middle of the last century which provided manmade substances of increasing sophistication, needed by fast-developing industries. Initially the new plastics were used as substitutes for natural materials which were fast being used up and this, coupled with unsuitable applications, gave them a downmarket image. But soon, new synthetics with special qualities began to be developed to suit the changing needs of industry – a process that is still continuing today, so that modern plastics have gained almost universal approval and respect.

The ever-widening line of products made by the synthetic plastics industry has been a major catalyst for crucial changes in industry in the twentieth century by making available plentiful supplies of reliable tailor-made raw materials. These are not only supplanting the old natural materials but even surpassing them; for instance, the entirely new materials with very special qualities like the heat-resisting plastics used to protect space vehicles and to coat nonstick skillets.

The ability of industry to make the major changes of the last few decades has not only depended on rapid development of technology but has also required new materials which are increasingly more specialized and available in consistent quality at an economic price. Existing "natural" materials such as beautiful woods from the rain-forests were fast running out and becoming expensive – and in any case were unstable, difficult to process, and hardly suited the new mechanized industry. What was needed therefore was a new line of raw-materials which could be specifically purpose-designed: an entirely new concept. The answer was of course plastics, which could be formed into solid objects of any desired shape and multiple characteristics such as being hard or soft, rigid or bendable, firm or elastic, slippery or frictional, light or heavy, dense or foamy, heat or cold resistant, transparent or opaque, plain or patterned, bullet-proof, shatterproof, water-proof, and in every possible desired color.

No natural plastics are capable of as many variations as these synthetics. Yet despite the undoubted beauty and usefulness of the manmade materials of the twentieth century, many of us still treasure the old natural materials like amber, silk, tortoiseshell, precious metals, gemstones, and even less exotic ones such as wood. In addition, the increasing rarity of such exhaustible resources of the earth makes them increasingly more valuable compared with new materials. If polyethylene were as rare as diamonds,

PLASKON
Molded Color
meets your prospects
more than half way!

"Attention!"...that's the command of products molded
from colorful Plaskon. They fairly leap into view, meeting buyers' eyes with a powerful, sales-stimulating force.

"Profits!"... that's the command of colorful Plaskon,
while developing production economies and increasing
output to meet growing sales demands.

Plaskon has become the world's largest-selling urea-formal-
dehyde plastic, because it so thoroughly meets today's needs
for effective, new, income-producing ideas in business.

Plaskon Molded Color is rich-looking, vibrant, vividly
toned. It can be supplied in a complete range of shades,
to harmonize with the character or design of any product
and to enhance appeal to all types of people. It has many
features such as lightness, strength, deep color and adapta-
bility to an endless variety of manufacturing requirements.

We can give you helpful assistance in the matter of
designs, qualified Plaskon molders, and technical advice
for fitting Plaskon Molded Color to your production
and selling program. A Plaskon representative will be
glad to call and give you complete details. Plaskon
Company, Inc., 2125 Sylvan Avenue, Toledo, Ohio.

PLASKON
TRADE MARK REGISTERED
MOLDED COLOR

LEFT Plaskon was one of many materials encompassed by the generic name "bakelite."

would it be seen as just as valuable? Plastics are mun-
dane by comparison, but because of their ability to be
transformed into a myriad of different shapes and
guises they have a magical quality all of their own.

As the French philosopher Roland Barthes has
observed: "It is the first magical substance which
consents to be prosaic. But it is precisely because this
prosaic character is a triumphant reason for its

Color sells!

BAKELITE AND RADIO

Bakelite was the first truly synthetic and entirely man-made plastic. It was produced in America by the "father of the plastics industry," Belgian-born Dr. Leo Baekeland, in 1907. This may surprise many people, since bakelite is widely thought of as an archetypal material of the Thirties. It actually took about 20 years for bakelite to establish itself as suitable for the manufacture of large articles, although Baekeland's promotion of it as "the material of a thousand uses" was quickly borne out.

After a spell of teaching at his old University of Ghent, he emigrated to America at the age of 26 and worked on photographic chemistry at the turn of the century, providing independent means for himself by selling his patent for a new photographic printing paper called Velox. It enabled him to pursue work on producing synthetic resins, following advances in the field by earlier researchers and experimenters. In 1907, after several years' work, he came up with the new substance he called "Bakelite," which was to make him world-famous. First used as lacquer and an impreg-

existence: for the first time, artifice aims at something common, not rare." He also reminds us that as well as convincing us that it can imitate the rare and beautiful, plastic has also climbed down and consented to produce the utilitarian and the commonplace: as well as providing us with convincing fake jewels, it also provide us with plastic buckets. Plastic also provides us with products that nature seems not to have perfected, useful materials with extreme applications and even unrottable spare parts for the human body. Yet plastics are curiously without real form in themselves, born out of messes in the bottom of test-tubes, they seem to have no life of their own, existing only in a kind of ephemeral, transitory way – and gaining reality only when transformed into something else by the technology of the modern alchemist.

nating fluid, bakelite was later combined with various fillers to create a Machine Age material capable of being used to turn out exact copies of a myriad of different articles from household and cosmetic goods to gear-wheels and radio cabinets, with great accuracy and speed.

When it first appeared on the market, this new material must have seemed the answer to the radio manufacturers' prayers. It promised an immediate end to the old, expensive, labor-intensive processes previously required to construct cabinets, enabling manufacturers to turn them out at speed like jellies from a mold. It was this unique quality which specially excited the burgeoning radio industry at the beginning of the Thirties, for it would replace the dozens of separate skilled operations that were required to make wooden cabinets by a single machine-operation, cutting the time required from perhaps an hour to a few minutes.

When it was established, bakelite found its way into almost everything: electrical components, car parts, toiletry packaging, furniture, kitchen equipment, children's toys, cameras, and even decorative objects for the home. It arrived just as the moment that the radio industry needed such a material to substitute for its existing unsuitable and expensive materials. Bakelite and the synthetic plastics which followed freed the designer from the restrictions inherent in the use of earlier, more rigid materials, letting them develop an entirely new line of possible forms.

The name "Bakelite" was also the trade-name of the company which first produced the new material commercially; but it soon became such a well-known and ubiquitous material that "bakelite" (with a small b) began to be used as a generic term – like Hoover and Biro – to describe a whole family of later products of similar kinds. Today, there is so much confusion in the popular mind about the many different forms of plastics now available – some of which are so difficult to distinguish – that objects made from them are commonly misnamed.

Some purists restrict the term "bakelite" strictly to Baekeland's original formula: phenol-formaldehyde resin and opaque fillers which are formed into powder and then injected into molds within heat-and-pressure machines to produce casts. But most enthusiasts would include under "bakelite" materials with a similar base known as "cast phenolic resins" – thick, colored,

ABOVE Dr Leo H. Baekeland, father of the modern plastics industry.

translucent syrups without "fillers" which require no pressure to form, being simply poured into molds, then oven-baked to harden them.

The latter method of casting radio cabinets was perfected by the Catalin Corporation of America, whose designers and craftsmen raised the radio cabinet to an art form in the Thirties. This was partly because of the somewhat dated "handmade" method of production, which reversed the de-skilling effects of the accelerating mass-production of the day by preserving a measure of skill on the part of the individual worker. An individual could demonstrate his artistry by mixing various color agents into his resin, then swirling the mixture to produce beautiful patterns that made each individual cabinet quite unique. The manufacturers of these "Catalin" radios also cleverly rang the changes on basic designs by casting cabinets in a number of sections cast in contrasting colors which could be mixed and matched to produce a large variety of combined patterns. Thus, the customer could choose, say, a basic

ABOVE Ekco's All-Electric model
AC97 was available either in walnut,
or in black and ivory.

ivory cabinet, but order it with a "trim" and knobs of another color; a popular mixture included red, white, and blue sections and was naturally dubbed the "Patriot" model. This was a neat extension of Henry Ford's economic sales-ploy, "You can have any color you like so long as it's black;" the customer certainly got his color choice but was supplied with exactly the same radio as everyone else got.

It was not until the late Twenties that bakelite was first used for moldings as large as radio cabinets, although its suitability for mass-producing a huge variety of small objects from toys and telephones to Ford car gear-wheels was seen from the very beginning. The reasons were twofold. Firstly, the capital costs of installing the gigantic presses and of making the large press-tools needed for bakelite molding were not within the means of the industry until it coalesced at around 1930 from a myriad of small firms into a group of very large and profitable ones. The second reason was resistance from potential customers to brighter colors and new shapes. They had been accustomed from the beginning of radio production to buying radios which matched the hand-me-down Victorian furniture in their stuffy parlors – cabinets made from "real" wood that looked like furniture. To counter the resistance, bakelite dyes were used to provide mottled effects in brown bakelite which gave a fair simulation of wood finishes; but potential customers were not really fooled. They wanted the cosy "warmth" and attractive "graining" that was unique to each individual real-wood cabinet, and claimed with some justification that natural wood made the radio sound better. Despite these objections bakelite soon became popular among less wealthy customers who were prepared to put up with these radios because they were cheaper. Even so, there continued to be a certain amount of resistance to bakelite radio cabinets, which had in some quarters acquired the "cheap and nasty" image. This unfair reputation had been earned by the inappropriate use of previous plastics as substitutes for natural materials in cheap, badly designed products some years earlier. In the case of radios, bakelite was seen as a substitute for the "real thing," precisely because it was often used to imitate wood.

ABOVE Wells Coates' Ekco model AD65 of 1934.

BRITISH ROUND EKCO RADIOS

The radio industry had begun to take an interest in bakelite in the early Twenties, employing it for electrical plugs, control-knobs, and small components used in the "works" of radios. But there seemed little thought of using it to make complete cabinets until the very beginning of the Thirties and even then, the first of the few all-bakelite cabinets, used for early battery-and-accumulator receivers, were very small.

A typical example of the way in which the history of the employment of bakelite as a material for radio cabinets developed is afforded by the British pioneers of synthetic plastic radio cabinets, E.K. Cole Limited, who marketed their sets under the name "Ekco." In 1930, after calamitous experiences with wooden cabinets for the mass production, they took the bold step of going over to bakelite. At first, they imported the plastic cabinets from Germany, which had already pioneered the development of very large moldings, but two years later they imported a complete molding-plant from the major German molders AEG (Allgemeine Elektrizitats Gesellschaft).

Initial sales of the first bakelite radios were disappointing, as many judged that a mere reproduction in plastic of a "wooden design" was unattractive and inferior. The result was that Ekco took another bold and innovative decision – they decided to abandon the old "furniture-style" cabinet designers on which the industry mostly depended, in favor of innovators of the Modern school of architecture. Such designers were interested in intelligently utilizing the unique qualities of the new plastics but also wanted to consider functional and production requirements, and to produce at the same time an attractive and saleable cabinet.

As a result of this policy Ekco met with remarkable success, although their "designer" cabinets at first appealed mainly to well-to-do people who saw themselves as "modern." Ordinary families still demanded nice pieces of furniture in real wood that would

EARLY BAKELITE MANUFACTURING

The British pioneers of synthetic plastic radio cabinets were E.K. Cole Limited. The company discovered the difficulties in mass production of wooden cabinets, and in 1930 took the bold step of changing to bakelite. Germany had already pioneered the development of large bakelite moldings. First E.K. Cole imported new cabinets from Germany, then a complete molding plant which they installed in their factory.

ABOVE A huge German machine, dwarfing the sweating operators, installed in Ekco's British factory at the beginning of the Thirties. It turned out AD65 radios under heat and pressure of more than 1000 tons.

LEFT Part-time workers. Women workers were used as cheap labor in the days before equal pay, cleaning up radios as they came off the press.

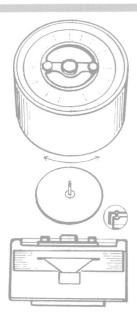

LEFT Sketch for an early round radio on a turntable, for use on a low table.

RIGHT In contrast to the heat-and-pressure technology needed for molding "filled" bakelite powders, the craftsman here is pouring Catalin – a phenolic liquid resin – into molds which are then baked in an oven to harden the cabinets before removal. This technique restored the artistic input of the craftsman, enabling him to produce jewel-like, translucent colors and beautiful marbleized effects.

ABOVE The famous British "Round Ekco" AD65 radio of the early Thirties, cast in brown bakelite.

LEFT Ekco's model AD36 of 1935, available to special order in a number of non-standard colors such as yellow and blue that are now considered incredibly rare.

go with the rest of the furnishings in their predominantly tobacco-brown and dun-cream homes. Proof of the point was that Ekco's first real bakelite "designer" radio – the revolutionary modern, circular AD65 model, conceived in stark, snazzy machine-age style black-and-chrome – was not half so popular then as it is among Art Deco collectors of today. The designer must have felt very frustrated when it failed to sell well – especially when it was recast in imitation mottled "walnut-wood" bakelite and sold like hot cakes at a slightly cheaper price. The same model was also offered in pale green imitation onyx and white marble – but nobody seemed to want these either.

Certain myths surround the AD65, or "Round Ekco" radio. This model, produced in 1934 but designed two years earlier, has repeatedly been described as "the very first completely circular radio" and the Art-Deco-style invention of British architect-designer Wells Coates.

Recent research has brought to light information which suggests that although this trail-blazing design was certainly the brilliant creation of Wells Coates, it is

also likely to have been partly the result of an evolutionary process in which a team of people took part – including marketing-men as well as designers – rather than the inspiration of one designer.

It seems that design proposals for an earlier and remarkably innovative circular radio were presented to a design meeting at the Ekco Radio works well before the AD65 was produced. Another ingenious architect-designer named Rodney Thomas, who worked for Wells Coates, presented to the meeting a sketch of a completely circular design which was intended to be used "flat" on its back on low tables. It was conceived as a battery-portable radio which integrated several "naturally" circular parts: a loudspeaker, a huge dial, a dual-concentric knob, an internal direction-finding aerial, and a turntable. All these were incorporated in a drum-shaped cabinet mounted on the turntable so that it could be rotated by hand to find the optimum direction of the desired broadcasting station's transmitter. It was a novel idea which exploited the then current "modern" preoccupation of "functionalism" – the marriage of form and function – rather than simply

RIGHT Wells Coates was probably influenced by the work of the architect Charles Holden, who was then incorporating drum-shapes in his designs for a series of London Transport underground stations.

an aesthetic expression of the Art Deco style which was being revived in avant-garde design circles. At the Ekco design-meeting, the proposal was received with interest and a suggestion was made that it could be fitted with a handle and marketed as a battery-portable, but that idea was rejected. Giving consideration to the design as a radio for the home situation, the Ekco sales people reckoned there would be no demand for a low-level radio because the tables in peoples' homes at the time would be too high to let listeners seated in armchairs look down on it to enable them to see the dial (low coffee tables had not at the time been introduced). However, they thought the round radio was such a nice shape that it would serve very well as the company's new electrically powered model if turned on its side. Alan Stewart of London, to whom this story was told by Rodney Thomas, says that Wells Coates protested that his design was reduced to "mere style" and comments: "A fine 'Modern' design was lost and a much-loved Art Deco design was born."

Wells Coates, who once said "A radio should never be disguised as anything else," was a well-established

and extremely imaginative architect. He designed various parts of the innovative Art Deco-style Broadcasting House in London for the British Broadcasting Corporation in 1932. His work exploited circular forms and he was probably influenced by the then current work of the architect Charles Holden, who was incorporating drum-shapes in his designs for a series of London Transport underground stations, notably the completely circular Southgate and Arnos Grove stations in North London (which are now listed as buildings of historic importance). Wells Coates was very much a man of his time rather than a hidebound classicist and must also have been influenced by the exciting new movements in art and design of the late Twenties and early Thirties which were challenging old and stagnant, square and classical institutional concepts to develop Modernism.

Other designers were working along similar lines to Wells Coates in various parts of the world – notably in the United States where there was a tremendous amount of innovation in radio cabinet design alongside the more traditional production of furniture-style sets.

With the impetus of the new bakelite-type plastics, they were able to produce radios in styles that seemed to precisely reflect the spirit of the Thirties, in bright new colors that made them function as decorative objects in their own right.

These designers of the Thirties were heavily influenced by the revivalist Art Deco movement, which had its roots in the revolutionary Modern Art movement of the Paris of the mid-Twenties when rebellious artists were producing geometric patterns and forms in startling colors, thrusting aside the stylized, natural forms and colors of earlier Art Nouveau. Into this

melting-pot of Art went Machine-Age concepts, cubism, constructivism, functionalism, kitsch, and many other new ideas derived from the traditional art-forms of Africa, the Aztecs, India, and Mexico: what came out was a completely new variety of styles of radio, although these took some time to become established.

By that time, broadcasting itself had changed too, from a rather serious system of more or less official communication and into a mass-medium which now also reflected the leisure and pleasure needs of people. The radio itself had become the leading machine for entertainment and information in the home, challenging the newspapers, the theaters, the movie-theaters, and the record industry – dramatically re-ordering the social life of everyone.

In its new and rapidly expanding role, the radio had suddenly taken over the central focus of everyone's home, upstaging the parlor piano and the gramophone with which they used to entertain themselves and giving them a broader view of the world. They could now hear the sounds of faraway places with strange sounding names, hear news as it happened, listen to a play or an opera or a church service, take part, in imagination, in a tea-dance at a swish hotel, or tune in to a live sports event. In addition to all this, the radio had also become a decorative object in its own right and would later be celebrated as an art form. It had enormous cultural effect too, generating an interest among people for all

kinds of activities they had not before been able to take part in. This also brought in new concepts of social status. The very fact that someone possessed the latest design of radio was enough to generate a competitive one-upmanship among neighbors and even the ostentatious display of a big radio aerial on the roof was an indicator of your forward-looking attitudes and thereby your position in the cultural pecking-order.

THE AESTHETICS OF PLASTICS

Unlike the Luddites of the First Industrial Revolution, who looked at the immediate practical problems of their contemporary situation and rejected the machine as a threat, the Modernists were optimistic, even Utopian, welcoming the new Machine Age as an answer to almost all problems and an opportunity to be rid of an established code of outdated aesthetics which was largely seen as mere justification for repeating what had gone before.

The Established cult of the "Beautiful" which had degenerated into a mere style of unnecessary and artificial imposition of classical forms and fussy decoration was soon to give way to a kind of severe plainness which anticipated the minimalism of the post-modern, partly because it was cheaper and easier to produce by machine, and partly due to a developing

phobia toward over-ornamental elaboration of pure mechanical forms and a requirement to become starkly different. This was celebrated as a kind of aesthetic "honesty" and led to a preoccupation among designers of making form fit function. They soon invented their own new code of aesthetics which would eventually degenerate, within the cult of Consumerism, into a conception of mere marketable style rather than the perfect combination of form and function they had originally craved.

One might with some justification observe that this development has continued to the present day, when the designer-object which fails to function as well as the old-fashioned implement it is meant to replace, is often greeted with amusement or derision, yet not actually rejected. These products often seem beautifully simple and user-friendly but may conceal unnecessarily sophisticated technology of mind-boggling complexity, giving the user assurance that he is being offered something thoroughly scientific although it is in fact just a fashion statement. And just as overelaboration, fussy detailing, overdecoration, and garish color was the fashion of Victorian times, so plainness, minimalism, fear of decoration, and absence of color has become a fashion among modern designers. These pre-occupations with style and fashion show just how much everyday products have always been endowed with meanings that have little to do with either art or utility:

LEFT The French–American company Sonora manufactured this late Forties Sonorette, influenced by the curvaceous styling found on American automobiles.

the objects are meant as a kind of cultural sign, indicating the status of the possessor.

The work of the more avant-garde artists and designers of the modernist period represented a conscious rejection of the rigid and sanitized aesthetics of the Victorians in favor of a less inhibited Art Brut which explored Sigmund Freud's newly defined Unconscious Mind as a creative force in art. This challenged the cosy respectability of the ordered forms and colors of the dying Arts and Crafts Movement and the Art Nouveau stylists, introducing rebellion that was both anarchic and democratic and attacking the very roots of questions such as "What constitutes Art?" and "Who is permitted to practise and produce Art?, to criticise and curate it?" It was followed by the jazzy Art Deco design movement in which designers seemed to be influenced directly by the rebel artists. The background to all this was the great relief of the mass of the people following the terrible 1914–1918 War, which the politicians pronounced "the war to end all wars," promising "a land fit for heroes" to come. Very much in the air, too, was a new faith in the promise of science and technology to create a better world.

Ironically, it was the new technology itself – in the form of mass-audience radio, television, and movies – which promoted and reinforced itself as the prime agent of hope for the future, emphasizing the advantages of modernity (like labor-saving home-machines and cheap automobiles) but failing to consider the down-side of science (such as the atomic bomb and gasoline-created pollution of the planet).

A comforting myth of salvation via science was born of a dreamscape desire in this Age of the Great Movies for everyone to aspire to the presumed wonderful lives of the stars as an escape from the miseries of the great economic depressions. Cults of new design in consumer goods grew out of the myth, like cars that looked like Dan Dare's spaceship and seemed to require highly technical wizards to drive them. Radio design was part of all this too – the radio became a symbol of modernity, reflecting the hope of science, the increasing technological literacy of the masses, the power of universal communication, and the instrument of escapist pleasure and entertainment.

There is also a kind of "Art" value in certain contemporary objects made from synthetic plastics which do not mimic other materials but which exploit the material for its own special characteristics, and create form and color according to its own vocabulary. The debate still goes on as to what degree form should submit to function or vice-versa – and objects continue to be produced which appear to be influenced in either way. The best designers engineer products which work well and also look good but bad ones continue to launch products which often amount to kitsch, bringing derision and contempt to the designer-label.

BELOW The influence of the automobile is obvious in this Spanish Artes AR3 receiver in sprayed bakelite.

Overemphasis of the outer form of objects often inhibits their functionality, but at the same time turns them into mere statements of style, and it is sometimes the case that an appearance of functionality is no guarantee that the product will suit its intended purpose – for functional appearance itself has become a style.

Since designers have increasingly become the tools of the marketing-men of Consumerism: their role is now seen to be one of packaging objects to sell rather than producing beautiful or functional things. Often, they are required to sell a dream rather than a real object: the designer who knows his psychology lives in a Walter Mitty world. Thus an Art Deco radio may transport the listener in his imagination back to a fantasy age of cocktails-and-laughter; while a technical-looking one of today with many knobs and dials may make him feel as though he has the technological know-how of a Concorde pilot.

In recent times, there has been a strange shift in the status-value of synthetic objects, demonstrating that status-value is not a fixed quality but a relative, self-defining, and continuous process of signification which has the function of letting particular social and economic groups of people distinguish themselves. For example, the first imitation fur coats, made from synthetic plastic fibers to mimic mink or fox, which were originally sold under euphemistic and misleading descriptions (as were the cheaper dyed cat and rabbit imitations) were initially rejected with distaste by people who could afford to buy real furs in order to show their superior status. But when the animal-rights movements began to jettison their "cranky" image and became socially respectable, the same coats were boldly renamed "fake furs" and were even dyed in vividly unnatural colors, subsequently finding a ready market among those who wished to be seen as humane and *au fait* with avant-garde fashions.

We have a continually changing perception of "The Plastics Age" in terms of synthetics which are now getting on for a century old. Plastics have by turns been celebrated and despised but they cannot fail to be enduring – at least for as long as the natural resources, from which they are derived, last us.

Perhaps ecologists can take comfort from the fact that many plastic objects can now be as long-lasting as those made from traditional materials, providing the juggernaut of consumerism can be persuaded to give up its giddy pace along the headlong course of built-in-obsolescence to a totally Throwaway Age.

At least some people are trying to halt this giddy pace and are preserving some of the artifacts of the past. In time, perhaps, today's classics in plastic will be seen to be as important as other prized antiques, and will be handed down from generation to generation as treasured objects or simply as enduring and useful things.

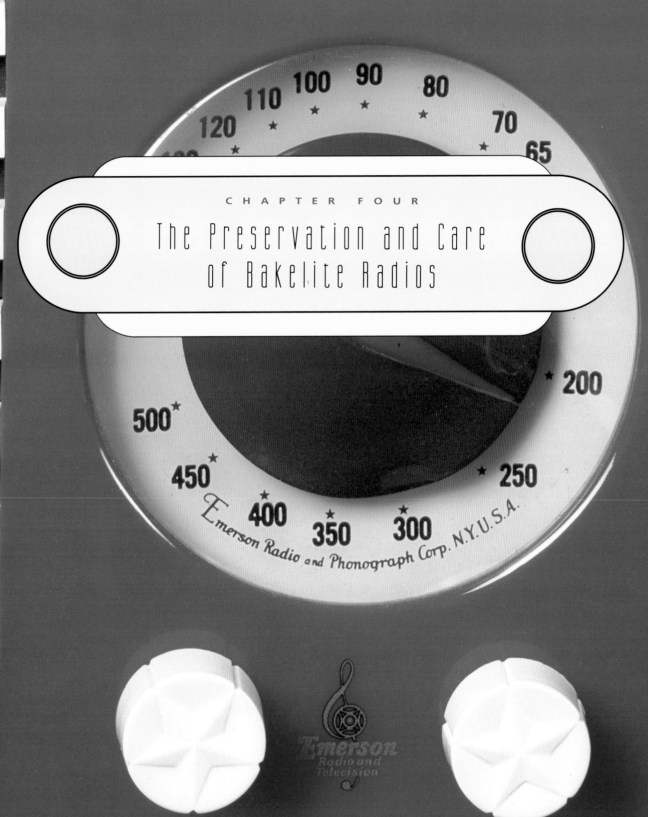

CHAPTER FOUR

The Preservation and Care
of Bakelite Radios

Why do people collect bakelite radios? There are many reasons, but most collectors are attracted by nostalgia, by the aesthetics of a particular period of design, by a fascination for the quaint and often highly inventive old technology, or simply by a desire to possess beautiful things.

Some antique radios have recently been sold for high prices and people are being told that they are an "investment." But if money is your only motivation for collecting, you are unlikely to get much fun out of it or make your fortune, for it's a risky business fraught with the problems of wildly fluctuating prices and fickle customers with changing tastes.

Some collectors are sparked off by inheriting a radio from their parents which they wish to preserve as a family heirloom. Luckily, our forebears bought their radios before the Throwaway Age began, when such things were made in such a way as to be repairable – unlike today's transistor sets with their solid-state works.

SOURCES OF SUPPLY

If you inherit an old radio you might find it interesting to collect some anecdotal information to go with it. Why not tape an interview with the owner who can tell you its history and comment on the old "Radio Days?"

There are many other ways of finding old radios. You could try advertising in your local papers or in specialized collectors' magazines – or you could put some cards on store notice-boards. You will need patience to follow up responses and may have to do a lot of telephoning and legwork but it could be worthwhile. Try to contact other collectors for help with repairs and restoration – or perhaps swapping activities. Perhaps you may be able to find a dedicated collectors' club which can give similar help.

Other possible sources of supply are street-markets, trunk sales, garage and loft sales, antique fairs, rummage sales, and junk stores; the ones out of town seem best. You need to arrive early since such places are now well-searched for radios. It's even worth looking into a builder's roadside garbage container.

If you buy in the streets, you may be just as likely to get a dud as a gem but will have no redress for a bad buy. Bear in mind that damaged or restored radios may be worth only a quarter of the price of perfect ones and a "mint" example in its original cardboard-box may fetch a high price – in fact, the box may be worth more than the radio because it is rarer! What dealers call "Provenance" is also valuable: documentation which proves pedigree such as a label bearing a dated message such as "The property of Winston Churchill" or "Dear Repairman, Please fix this radio, signed Elvis Presley."

A simple way to acquire old radios is to keep an eye on auction houses which deal in such "antiques." You will need to get catalogs of sales and attend viewings to examine carefully what is on offer, since the principle of *caveat emptor* ("Let the buyer beware") applies. Even if you do not buy anything, it is an advantage to know prices, to get experience of all kinds of radios, and to learn the tricks of the trade. Learn about bidding methods too: the auctioneer may begin by going up in single dollars to ten, then switch to fives or tens, so if you are an excitable person, you may end up paying more than you intended – and there will be an auctioneer's commission and perhaps cash to be paid on top of the hammer price. Make up your mind

ABOVE If a once-rare radio begins to turn up in greater numbers, such as this Czechoslovakian Tesla Talisman of the late Forties which is easier to find now than in Iron Curtain days, its value will drop considerably.

BELOW A radio still packaged in its original cardboard-box may fetch a high price. (Radio pictured on page 86.)

beforehand on a top figure, then stop when it is reached. Stories you hear about people accidentally bidding by inadvertently scratching their heads are largely myth, but if a gesture of yours is mistaken for a bid, you must speak up immediately.

You also need to learn some tricks of the trade. For instance, if people are dismissive about a "lot" at the viewing, they may be trying to put you off bidding for the item so that they can get it cheaply themselves. Beware also of viewers who "talk up" an item because they may be trying to encourage you to buy a dud lot they have put in which might be easier to sell in an auction than in their store. You buy goods "as seen" and have no redress if you get a bad bargain – except if you can prove there was deliberate deception on the part of the seller. Do not rely on catalog descriptions or estimates and remember there have been cases in which people enter lots which they hype up by arranging ghost bidders at the sale; then buy back the items themselves, thus establishing an inflated "going price" for such items. You should also beware of the illegal "Ring": a group of buyers who agree on the lots they will each bid for, insuring they will not force up the prices by competing with each other, then secretly share the spoils when the sale is over.

Finally, there is the alternative of buying from a dealer. There are, of course, good and bad ones, so check them out with fellow collectors, asking whether they have a reputation for straightforward deals, honest descriptions, and money-back guarantees. Buying from a dealer can bring advantages: he can help you find scarce items and build up your collection, take part-exchanges, and give advice. It may be worth paying a higher price for such service.

Whichever way you go about acquiring your radios, you will get more fun out of your hobby if you do some research in libraries and museums to enrich your historical knowledge and put your collection into a social context. In addition, practical knowledge will assist your buying and enable you to carry out repairs and restoration work.

BUYING BAKELITE RADIOS: WHAT TO LOOK FOR

Often, you may be unable to make a thorough check of a prospective purchase, but there are skills you can gain from experience that will enable you to spot faults quickly. Learn to recognize bakelite-type plastics in all their forms: they can easily be confused with later plastics and such knowledge is a good way of dating radios.

Bakelite moldings that have been formed by heat-and-pressure are rigid, brittle, heavy, fairly thick, and have a hard feel, are always opaque and could only be made in dull colors: mostly brown or black, less often in dark reds and greens. They can be plain or mottled to simulate mahogany or walnut. These colors are not just on the surface like paint, so look inside to check that

LEFT British Bush DAC90s are a fairly common find, and are popular and affordable.

INSIDE RADIOS

From the original crystal set to the modern day receivers, the works inside radios have become increasingly smaller. The British Lissen Crystal set, below, is a far cry from today's solid-state radios, with their machine assembled parts and microchip technology.

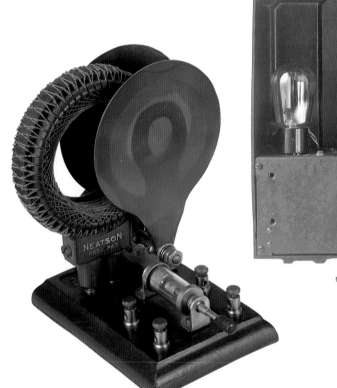

ABOVE The simple works of this 1925 British Lissen crystal set are all visible on the surface: crystal-detector, coil and tuning-device. It needed only an aerial and a pair of earphones to work and required no electricity since it was powered entirely by the incoming signal from the broadcasting station.

ABOVE The inside of a French Thirties bakelite radio, showing its valves and internal loudspeaker. It also needed an aerial and was plugged into electric power.

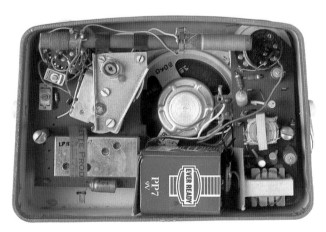

LEFT The inside of an early transistor portable radio of just after World War II, showing tiny transistors which replaced hot, power-hungry tubes and consumed only a fraction of the electricity. Each of the tiny parts had to be separately soldered into place. The aerial was internal and it worked on batteries.

RIGHT A Motorola, model 50XC, made in 1940.

the color goes all the way through. Occasionally, ordinary brown bakelite was used, then sprayed over with brighter colors but not all such painted sets are genuine: painting sometimes indicates that recent repairs have been concealed.

"Catalin" type cabinets which are made by pouring resin mixtures into molds have a softer, more flexible feel about them and are more easily cracked, worn, and scratched than hard bakelite ones. They are also more translucent and can be in bright, jewel-like hues which may change color with prolonged exposure to ultra-violet light and will distort if subjected to heat and sunlight. Blues turn to dark greens with age and white goes yellow but this can be corrected by an expert. Often, tops have burn marks caused by the hot valves inside which cannot be removed.

The Catalin-type cabinets are perhaps more easily confused with modern plastics by the novice but when you have become familiar with them, you will find them unmistakable and will be able to spot fakes.

Check that knobs, plastic and chromium decoration pieces, dials, and glasses are original, as these are being reproduced, although you may find them more acceptable than non-matching ones from other sets.

In judging a prospective purchase, try to inspect the inside too for repair patches and signs of corrosion due to dampness. And use a trained nose as well as your expert eye to check for signs of overheating that

could mean an expensive electrical overhaul. Examine the cloth covering the loudspeaker aperture: if it doesn't look in the same condition as the rest of the set, it may be a replacement but this may be acceptable if a good match to an original one.

PROTECTING YOUR COLLECTION

Since bakelite radios are now becoming expensive, you would be advised to insure your collection and protect it with some security measures. Some Art Deco radios cost several thousand dollars apiece. At a London auction not long ago, a rare "Round Ekco" in imitation green marble – which puzzled some experts – fetched almost £20,000, which compares with about £500 for an identical plain brown one. Such freak results rarely set precedents; if another appears, it is likely to fetch a great deal less, particularly since such events spark off a spate of reproductions making potential buyers wary.

But what is a radio worth? That, as any dealer will tell you, is simply a matter of how much collectors are willing to pay. Prices are mostly established at auctions and by dealers who are able to charge "what the market will bear." There is an approximate standard price for many radios which is based on "good condition" but "mint" examples may fetch a third more and "poor" but still original and complete ones only half that, while cracked, broken or incomplete ones may

RIGHT A postwar Bullet, a Catalin model very sought after in blue.

be unsaleable except to pirate for spare parts. Changing fashions and fads also cause price fluctuations among fickle collectors who like to change their love-objects often, so today's bargains can easily become tomorrow's worst buy.

You should have your collection properly valued for insurance purposes. If you are a good customer, your radio dealer may help you with this – but you should not necessarily expect an expert to give you the benefit of his knowledge free of charge. Auctioneers will give free valuations of items you may intend to put up for auction but these may not match store prices or "insurance value" which takes into account the price of a rapid replacement. Don't underinsure, but there is no point in overinsuring either, since you will only be paid the "current market value" at the time of replacement.

It is a good idea to put "invisible-ink" security markings on things to help identify them if stolen and to photograph them, writing details on the back of the prints to build up a pocket album of your collection.

BEWARE! DON'T PLUG IT IN

After you have made your purchase – what next? Do you immediately plug the radio into an outlet and wait for it to blow a fuse, or possibly burst into flames if you have not checked the voltage is the same as your supply? The answer is "No." If you are not technically

minded, do not fiddle with it or you could get a dangerous shock, or indeed blow it up. Get an expert to check it first. If the set works, you should still treat it with care: do not let young children near it and do not operate it in damp places like kitchens and bathrooms because it is highly unlikely to conform to today's safety standards.

There are other questions you should ask yourself. Is the radio a battery set? Are you still able to obtain batteries? What voltage does it require? Does it work on AC or DC? What does "60 cycles" mean? Does it require an aerial? Should it be earthed? If you do not know the answers, take advice from a reliable source rather than take risks.

Wherever you find your radios, it is important to bear in mind that many countries now have tough safety laws concerning electrical items which forbid the sale and use of equipment that does not comply with modern safety standards. It is unlikely that any vintage radios – except battery-operated ones – will comply with such regulations, so responsible professionals who trade in them are likely to take the necessary precautions. They may render the radio inoperable by removing the electric cord or other components; or they may adapt the set to make it safe to use; or may simply snip off the plug and sell it as "antique apparatus for display and research only" (the latter strategy may not protect them from the law).

BELOW The Fisk Radiolette.

WILL YOUR RADIO WORK?

When you buy a vintage bakelite radio, you should remember that broadcasting has changed very much since it was designed. Old radios will usually only tune-in programmes broadcast on the original "AM" (Amplitude Modulation) system on "Long," "Medium," and "Short" wavebands which are often shown on the dial of the set. These receivers will not tune-in stations broadcasting on the FM (Frequency Modulation) system introduced in recent times – which you will not find marked on the dials of early receivers. Sometimes, particular stations may at present be broadcasting on the old AM systems as well as FM and in this case you may be lucky enough to be able to get your favorite program but if not, you will have to put up with whatever is available. It is not feasible to convert such sets to FM but you might be able to get a separate converter. Alternatively, you could feed the set with music to match its age from a hidden cassette-recorder with the help of an electronic engineer: the trick with this is to tell your friends you have bought an old radio that still gets old sounds and to astonish them with a demonstration.

CARE AND REPAIR

If you need to repair or restore your radio, take care or you may ruin it. Museums aim to conserve and preserve rather than to restore or repair objects. They are concerned with keeping things in safe conditions and adopt the policy, "Do nothing that cannot be reversed," so if it is decided to deal with cabinet damage for instance, removable wax rather than anything permanent would be used. Any changes made are recorded on a label for future historians. You may not wish to be so strict but even so, your maxim should be "If in doubt – leave it alone." If you decide to do repairs and restoration, take some safety precautions before starting work. Wear protective clothing and use a face-mask and goggles if using harmful solvents, abrasives, and electric tools.

If the radio is undamaged and works, it may simply need a gentle cleaning and for this you can use a mild detergent solution followed by a good wax polish – but do not get water inside it and make sure it is unplugged before starting work. If you need to remove the works, remove screws carefully and make a note of where everything goes for reassembly.

On display, some cabinets seem to attract dust, so an antistatic polish may help. Brownish-yellow discoloration often results from a sticky deposit of tars from tobacco smoke which can be removed with detergent; but if you are a smoker, you may have already considered giving it up to finance your new collecting habit.

A set which is badly soiled may require the use of special abrasive polishes but in this case begin with the most gentle kind because shiny surfaces are often a very thin skin which if broken may expose a rough core that can never be polished. Paint spots can be removed with a plastic scraper or diluted chemical stripper which should be first tested for corrosiveness on an unseen area. Cracks can be filled with colored wax of the sort used by furniture restorers and broken off pieces can be secured with an epoxy resin glue. If you possess the sort of skills which car repairers have, you will be able to fill in missing areas with material used in their trade – but you will need to dye it to match first. Experts use cold-casting resins to mold missing parts, like knobs, dial

BELOW A 1940 Catalin radio designed for Emerson.

surrounds, and decorative appliqué, some of which can be bought ready-made. Different techniques are needed according to the type of cabinet material, the most difficult of which to repair are the Catalin types which are translucent, so almost any repair will show.

The acquisition, study, restoration, care, and usage of bakelite radios are important matters, whatever the purpose of the collector: whether a hobbyist, home-maker, aesthete, technician, designer, historian, or museum-curator. But further studies which consider the objects in their social, psychological, and domestic contexts will greatly enrich the enjoyment and importance of the collection of any radiophile.

POSTSCRIPT

In the Golden Age of Radio the domestication of a new and tremendously important communicative technology transformed the private space of people's lives. The radio in the home connected the private with the public, the old with the new, the present with the past and with the imminent future. Radio has led the way to the marvels of a new Electronic Age, yet all this apparently alien technology rapidly becomes part of the domestic scene – just like the bakelite radios of the Golden Age.

Synthetic plastics, of which bakelite was the pioneer, have changed almost every aspect of our lives, just as communications technology has revolutionized the way we experience the world. The bakelite radio acts as a kind of milestone symbolizing these develop-ments of modernity. Plastics, once looked upon as cheap-and-nasty substitutes for the real thing, have become thoroughly respectable – we have even installed them inside our bodies as heart-valves powered with electronics, to renew us with the very power of life. The experience of modernity has taken us nearer to the stars but not quite yet to Heaven: unless it may be the result of a final atomic explosion.

ROBERT HAWES

CHAPTER FIVE

Collector's Directory

The following collector's directory shows a selection of Bakelite radios demonstrating the remarkable variety produced in many countries from the mid-Twenties until a few years after World War II. They range from the early, stylish but restrained Art Deco designs to the extravagant and even outrageous objects that marked the sad but triumphant end of the Golden Age of Radio.

It was the introduction of plastics, beginning with Bakelite, which freed cabinet designers from the constraints that had been imposed on their creative endeavors by the necessity of employing the traditional natural materials – mainly wood – which produced furniture-style cabinets intended to blend with the interior accoutrements of the typical home of the Thirties. New design movements like that of the Bauhaus, as well as influences from modern art, were changing architecture and interiors which now needed

equipment to match, so innovative designers began to produce furniture to match, which meant that the old wooden-box radio became an anachronism in fashionable homes. The immense design possibilities of plastics as well as its ability to match mass-production requirements was rapidly to change the look of the radio. At first, the new avant-garde designs appealed only to well-to-do, style-conscious people but were resisted by ordinary homemakers who could not afford to change everything instantly to keep up with the whims of the stylists; for them, the Art Deco bakelite radio would have been an impudent anachronism

in the living-room. Consequently, the more extreme upmarket designs were diluted for the mass-market until general furnishing style came up to date, the chagrin of designers who saw the innovative shapes and bright colors of their creations debased into more traditional styles which imitated dark-brown wood. But for many reasons, particularly the increasing cost of rare woods from the already plundered and depleted rain-forests, plastics were here to stay. Radios were to become part of the Machine Age and their form reflected the spirit of the age, symbolizing the middle of a Depression, the jazzy lifestyles of high society, and the exciting promise of automated technology to free working people from drudgery and unemployment. Radios took the forms of the monolithic, soaring sky-scraper; the streamlined airplane; the thundering locomotive and the kitsch automobile, reflecting all of the vibrant forms and colors of Modern Art.

The directory reflects all these developments, which began as small movements but rapidly spread all over the world. We have selected examples from the United States, Great Britain, France, Germany, Australia, Spain, Italy, and Czechoslovakia – all of which stamped an individuality on their designs.

AMERICA

Every shape, style, color, and material was utilized to satisfy the varied tastes of the mass-consumer market of America, with strong preferences prevalent in the diverse markets of the Eastern, Western, Southern, and Central States. Nowadays, in the ceaseless search for the past, the hobby of radio collecting is extremely well developed, whether you collect a certain material, subject, style, or manufacturer – the choice is nearly limitless. Unique to North America, the Catalin, cast phenolic resin cabinet, with its vibrant, solid and marbleized colors emulating rich, semi-precious stones, has resulted in an unparalleled lust for this particular genre of radios.

RIGHT

You would think that a radio called the Tombstone was some gargantuan megalith. It is actually a midget radio only 10 inches in height. Its early Gothic look harks back to its wooden ancestors, and is a strong indication of its position as one of Catalin's first cabinets. Made by Emerson in 1937, the model AU 190 is now highly prized in any of its dramatically marbleized colors.

RIGHT

One company producing Catalin radios in Canada was Addison. This is their rather up-market version from 1940, the model 5. This monumental radio can be found in three striking, sometimes heavily marbleized colors: yellow, dark green and red.

BELOW

The Little Miracle manufactured by Emerson in 1938 is another example of early cabinet design. The body was molded in a multitude of attractive, vibrant colors, while its grille and knobs were injection molded – through the years these have become prone to shrinkage, warping, and ultimately disintegration!

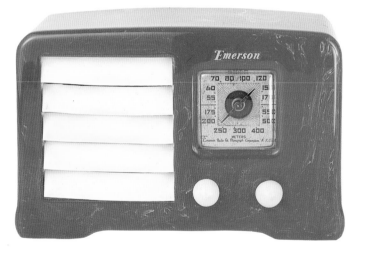

ABOVE

With its three interlocking Olympic style circles, this scarce and unusual radio was manufactured by Garod in 1940. This particular radio was marketed under the AMC name.

A chunky automobile-fender-grilled radio, whose chassis is unusually mounted upside-down in its cabinet. Made by Sentinel in 1945, it is particularly stunning and desirable in oxblood red.

Ovoid in shape and known as the Temple, this Fada radio came in two very similar models, pre- and postwar. Although not rare, they would certainly be an integral part of any Catalin radio collection.

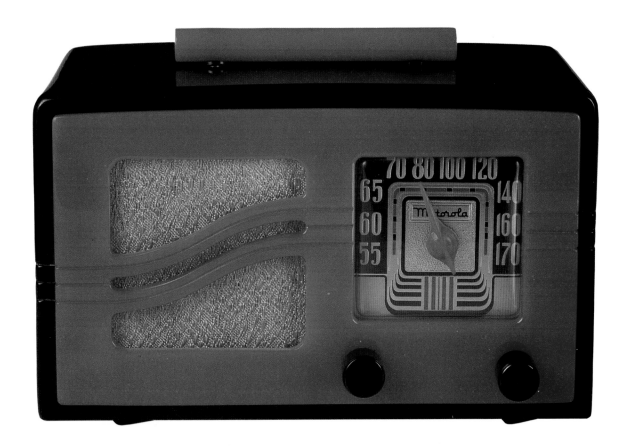

A delightfully solid and striking Art Deco design that was the last of Motorola's classic Catalin radios. Its sleek S-shaped front panel and striking dial face make this one of the most sought, and consequently rare, radios around!

LEFT

This radio was designed by Norman Bel Geddes in 1940 to celebrate the 25th anniversary of Emerson Radio. Known as the Patriot, it paid obvious homage to the star-spangled banner with its many combinations of red, white, and blue.

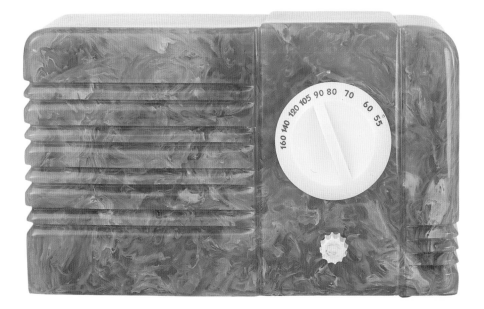

ABOVE

A solid green marbleized lump of Catalin, this small and chunky RCA radio known as the Little Nipper is unusual in that it has no grille cloth, no celluloid dial face, and no conspicuous numbers.

LEFT

This quirky Kadette looks more like a clock than a radio. The Clockette of 1937 was produced in a range of five stunning, luminescent "Crystlin" colors, the translucent glass-look blue being both the scarcest and most attractive.

ABOVE

The General Electric aptly named Jewel Box with its lavishly generous, thick, marbleized, sometimes translucent red case, and bulbous yellow knobs, is a joy to look at and touch.

Brittle bakelite and metal as well as a painted, usually worn map insure that an undamaged version of this novel, 1934 Colonial, New World Globe radio is very hard to find.

ABOVE

Designed by Harold Van Doren,
three quite similar models of this
skyscraper-influenced radio were
produced by Air King Industries in a
large array of unusually vibrant
colors. Until 1933, it was the largest
US bakelite molding produced – its
spectacular, monumental features
were possibly never outdone by any
other radio. Extremely rare,
especially in red,
blue or green.

Fada produced a number of similar, small, colorful cabinets early on. Of these the model 188, introduced in November 1940 for Christmas, is surely the scarcest, with only a handful known to exist. Known as the All-American, it was Fada's answer to radios such as Emerson's Patriot, with its red wrap-around grille, blue knobs, handle, and originally brilliant white cabinet.

ABOVE

Known as the Bullet, this classic streamline was one of the most popular and successful radios ever made. Like the Temple, it came in both a prewar gumdrop knob (top) and postwar fluted knob (below) model. In their many attractive color combinations they are very popular, and are normally a radio collector's first Catalin choice.

LEFT

The model 52 of 1939 was Motorola's first venture into Catalin radios. Its subtle indented geometric lines, crisp silver-foiled dial, and contrasting graduating speaker grille make this a striking and desirable radio.

RIGHT

The "Circle Grille" of 1940 came in a multitude of unusual Catalin colors from this rich honey example to a water-melon red or luxurious emerald green. As with all of Motorola's models, the 50XC is considered both rare and very collectible.

BELOW

The pre-war Garod of 1940, although quite large, is both attractive and somewhat ingenious with its unique, flush-fitting, hideaway drop handle. Another scarcity, this particular color combination is the only one known.

This Motorola of 1950 has obvious automobile influences. Designed by Jean Otis Reinecke in various hues of brown and cream, it epitomizes the feel of the Fifties.

BELOW

This unusual radio with its Roosevelt-inspired Scottie Dog motif and geometric Art Deco design was one of many similar radios manufactured in small quantities by Remler of Los Angeles in 1936.

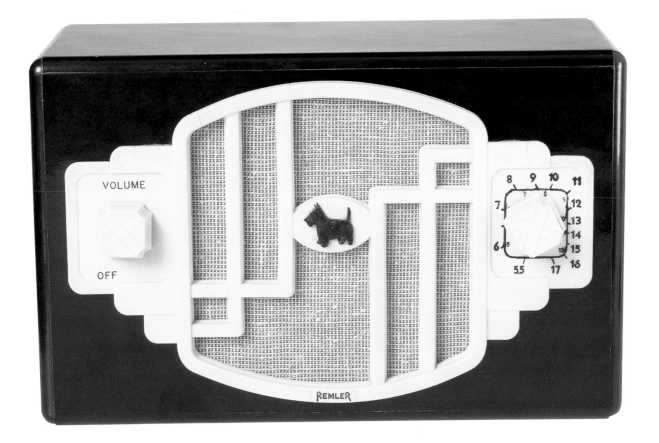

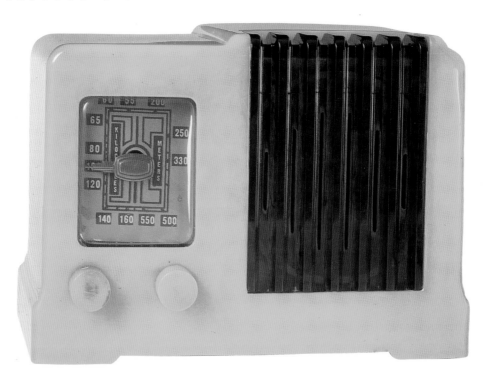

ABOVE

ABOVE

The subtly-stepped Globe radio of 1938 breaks the unwritten rule of putting the speaker to the left. In its case, its speaker is placed to the right. Nearly identical to the Arvin, this radio, with its vertically ribbed inset grille, is quirky in design, and much sought.

ABOVE

This Symphony, manufactured by Record-O-Vox Inc., is one of only two known examples. The color combination of emerald green and butterscotch make it what radio collectors' dreams are made of!

RIGHT

This baby Air King with its anodized metal grille seems to bear close relation to its larger, more well known architectural relative. It is made of Plaskon, one of the many materials to be now encompassed by the generic name "bakelite."

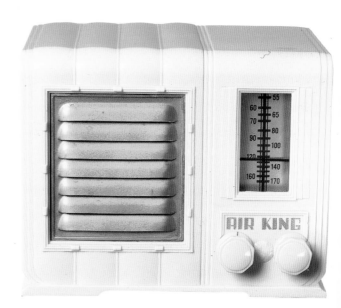

The Studebaker-inspired Crosley of 1951 came in a multitude of sometimes striking metallic and non-metallic colors sprayed onto bakelite. Like a number of other radios of the period, it is evidently influenced by both the grille and dashboard of the automobile.

The owl-like Zenith Crest of 1952 displays ingenuity of design in its small size and integral frame aerial which is neatly concealed in its liftable handle. Made in a multitude of different spray-painted bakelite colors, it is still underrated.

With its wonderful cascading waterfall grille, the Canadian Addison of 1940 came in a large variety of solid and marbleized color combinations. Made in either Catalin or Plaskon, this gorgeous red and yellow Catalin example is one of the most sought color combinations.

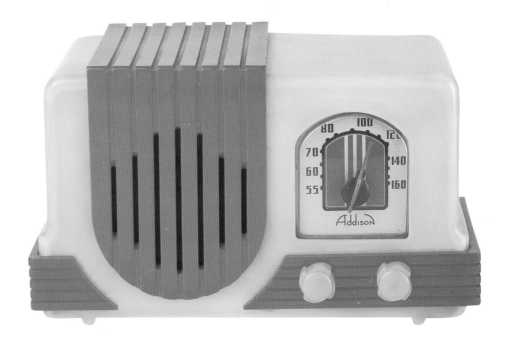

Although relatively common, the Bendix 526C of 1946 is certainly one of the most robustly constructed radios ever made. Its color combinations of green and black as well as its shape are striking and unique in Catalin radio design.

RIGHT

The Fada Coloradio is a glorious example of the Golden Age of Radio with its colorful red Plaskon cabinet and exuberant geometrical chrome tin. The model 254 came in various combinations, with this Chinese Red version of 1937 then being the most expensive. Nowadays, red is both desirable and scarce, making this radio two to three times more valuable than its nearest relative.

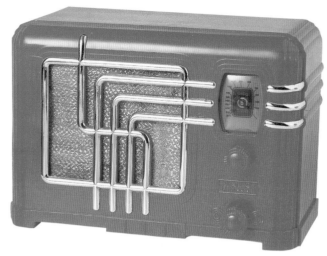

LEFT

The Emerson model U5A miniature Tombstone was a cheaper alternative to its internally identical, luxuriously marbleized Catalin cousin, the Emerson AU190. This radio was manufactured from bakelite or Plaskon which seems to have become prone to some crazing through the years.

RIGHT

An extremely small midget radio, this Detrola Super Pee Wee of 1938 should be considered rare in its blue veined, Plaskon form.

LEFT AND BELOW

"The World's Smallest, Power Packed, AC-DC Superheterodyne!" This was Emerson's claim for its Miniature Miracle of 1947. This pea-green example was found in mint condition inside its original box.

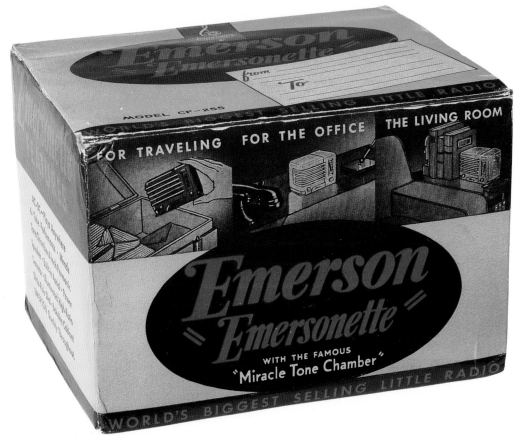

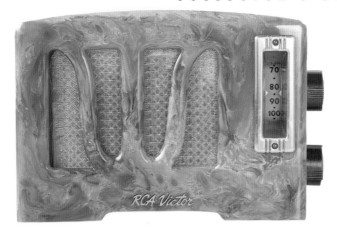

RCA's model RC350, a midget in the true sense of the word, measured a mere 7 x 4¾ x 4¼ inches! Especially attractive in this rare marbleized green version, its front grille area with its cut-out tulip design is unusual.

RIGHT

Aptly named the Kadette Jewel, this mid-Thirties midget radio with its contrasting marbleized fretwork grille and brass escutcheoned knobs is a sensational example of ornate Art Deco style.

RIGHT

Crosley produced this split grille, deeply grooved, knight-bedecked beauty in 1938. Manufactured both in the United States and Canada, it must be considered rare in any of its three color combinations.

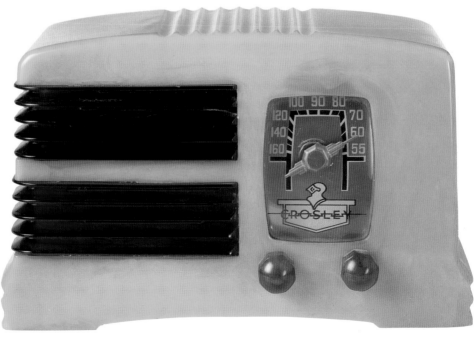

This rather quirky and off-beat radio was produced by Emerson in 1937. It seems like a forerunner to the more popular and now highly prized Little Miracle. Although the BN258 is not sought, it is extremely rare, specially examples such as this, with its unusual gold trim.

RIGHT

The cut-out grille, model 771 Air King, although having the same outline as its much more illustrious relative, is totally different. Its speaker is mounted to the front behind a large expanse of cloth, its dial is far larger, and it has the unusual feature of a crowned, "Magic Eye" tuning indicator.

ABOVE

Its offset pointer, large flat expanse of Catalin, and unusual color make this Sentinel of 1939 stand out. This model is both rare and desirable, specially intact as it has a very great tendency to crack.

BELOW

Although Zenith was a major player in the production of radios, it never moved very deeply into bakelite cabinet production. The Wavemagnet of 1938 is probably its best, with its curvaceous streamlined design, Art Deco dial, and incredible jutting out tuning knob.

ABOVE

An assortment of beer, Pepsi Cola, and champagne labels were but some of those attached to these uniquely shaped bottle radios. Made of bakelite and standing 3 feet tall, the bottle stopper acted as the tuning mechanism while the speaker was mounted upside down at its base.

This Canadian Crosley of 1953 is both a radio and a wake-up alarm. Made of bakelite coated with white paint, the model D-25 is yet another example of the automotive styling that predominated in Fifties' radio design.

Its striking, contrasting cherry-red grille makes this early Fada of 1938 much sought. Usually suffering from tube burn, the model 53X is specially difficult to find in pristine condition due to the proximity of the tubes to the side of the cabinet.

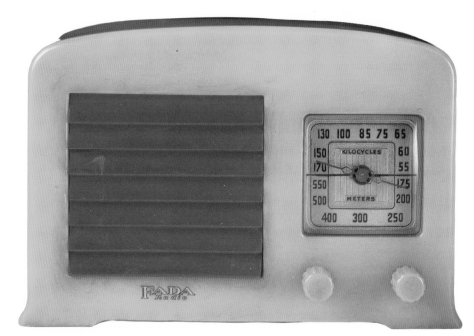

Though not the most exciting of Catalin radio designs, the EP375, known as the Five Plus One, has the unusually inspired feature of having six vertical Cellulose Acetate bars. It was produced in 1941 in a variety of unconventional colors, sometimes with the inclusion of brass rivets or integral handles.

BELOW

The postwar Fada Bullet was on occasion produced with a separate, contrasting inset grille. More sought than its straightforward compatriot, examples are now exceedingly rare and desirable.

GREAT BRITAIN

Although affected by the war years, radio production in Great Britain was both extensive and inventive. Beginning in the early Thirties, production mainly consisted of large conservative brown or black cabinets with the emphasis on design rather than color. The Wells Coates designed, ground-breaking Round Ekco of 1934 represented Britain at its most daring. The growth in the number of collectors in Great Britain over the past few years has insured that good, original examples of classic design can verge into the realms of the expensive, and can certainly be hard to find.

ABOVE

This small Goltone Super crystal set of 1925 was probably the first ever bakelite receiver.

LEFT

Looking like the front of a streamlined locomotive, the Philco Empire Automatic of 1938 was advertised originally as "the radio even your wife could use!" This luxury, push-button radio was pre-programable by any handy husband! It is now quite rare due to the weak construction of its push-button system which quickly made it redundant and replaceable.

The Lissen of 1931 was one of the few radios produced to have the Egyptian Art Deco look. This well-proportioned radio is fragile and thus rare today.

This "free gift" in return for 500 Best Dark Virginia cigarette coupons was one of the first bakelite radios produced in Britain in 1930. Manufactured by Kolster-Brandes Ltd., you needed to spend a small fortune and virtually smoke yourself to death to secure this 2-tube model called the Masterpiece.

ABOVE

One of the few skyscraper radios produced in the UK, this GEC Universal Mains Three of 1934 epitomizes the American culture of its day.

ABOVE

Bakelite really did release the designer's hand. For the first time in Britain an architect, Wells Coates, revolutionized the design of the radio with his Ekco model AD65 of 1934. Known as the Round Ekco, it was the first of five very successful models produced in either walnut or black bakelite with chrome-plated trim. Non-standard colors such as a marbleized green version were available to special order, these are now probably the most sought and prized radios to be found.

RIGHT

A radio price-fixing agreement prevented the supply of receivers to Co-op stores in Britain because the "dividend" they offered was seen as unfair price-cutting. The Co-op, in response, "defiantly" produced their own radio, the Defiant. The M900 of 1935 with its unique styling is now one of the most desired of all English radios.

ABOVE

The Ekco model AD75 brought out
in 1940 was designed to meet
wartime needs. "By exercising
economy in design, the rising costs
of the components have been offset,
but no attempt has been made to
cheapen either the materials or
the finish."

More Bush DAC90s seem to have
been made than any other bakelite
radio. It was available from 1946
well into the 1950s in a variety of
models in either brown, black or
cream. Although quite common,
their straightforward minimalist
design makes them both popular
and affordable, and they are
snatched-up by many aspiring
radio collectors.

Shown in 1946 at the "Britain Can Make It" Exhibition, this burgundy Murphy model A100 radio shows an innovative and fresh look at compact, user-friendly, modern design.

A radio designed with economy and thought, this Kolster Brandes BM20 of 1950 was made of two exact same halves that were then bolted together. Manufactured in a plethora of sometimes quite unique speckled and solid colors, they are now considered a subject for collecting all on their own.

RIGHT

Known as the Toaster, this Kolster
Brandes midget radio of 1950
generally had a cream, spray-
painted bakelite case. Other colors
such as this red one are rare
and collectible.

ABOVE

The ham-can-shaped Philips Local
Station Receiver of 1931 was made
of a material called Arbolite, made
by impregnating a sandwich of
fiberboard, woodgrain-finish paper
and transparent film with bakelite

resin under heat and pressure.
Simulating rosewood, it was used on
a number of early Philips' radios.
Due to the fragility of Arbolite, it is
hard to find the model 930A in
pristine condition.

ABOVE

Heptagonal in shape, this early
Philips Radioplayer of 1931 with its
oxidized bronze grille is much
sought for its outlandish and unique
appearance.

First produced on August 17, 1931, the Ekco All-Electric Consolette, model RS3, was the first British receiver to have a full range of interchangeable station names printed on the dial. Designed by J. K. White, its appearance, with an anodized copper grille, is synonymous with wireless design of the Thirties.

RIGHT

Fragile and possibly unique, this cream Ekco AC85 of 1934 is only one of a handful of known, "made to special order" colored models produced by the Southend-on-Sea factory of E. K. Cole. The instability of urea formaldehyde when molding large expanses has, over the years, caused a number of stress cracks to appear. "Colored Ekcos" must nowadays be viewed with great care and suspicion as their substantial value makes reproduction, for some people, a great temptation.

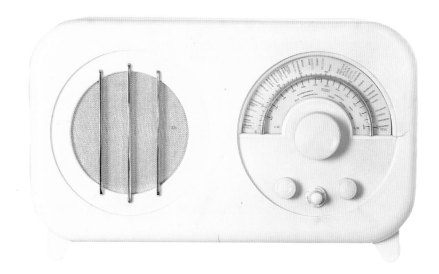

RIGHT

Designed by Jesse Collins, Ekco's tall, architecturally inspired All-Electric model AC97 had the added feature of a "Magic-Eye" tuning indicator. It let the user accurately tune in the radio by causing both halves of the green, luminescent, Xenon gas "Eye" to meet. This magnificent radio was available either in walnut, or for 10s.6d more in an incredible black and ivory.

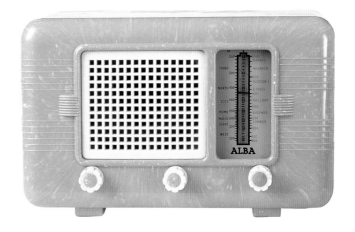

LEFT

This midget, Midgetronic radio of 1950 came in a number of color combinations, some sprayed on, others solid. In 1953 the cabinet, knobs, and dial were sold separately to enable home-constructors to fit their own chassis.

ABOVE

Manufactured by A. J. Balcombe Ltd., the Alba model C112 of 1947 came in a variety of interesting pastel shades. With its inlaid flower-shaped knobs and contrasting speaker grille, it can be considered collectible, specially in its strongest colors, such as green and blue.

RIGHT

The scale on the Ekco AC64 of 1933 was made adjustable by an interchangeable celluloid dial provided by the manufacturer. New wavelength allocations could be inserted and fixed with studs over the permanent meter scale.

LEFT

Designed by Eden Minns, the
Murphy model AD94 with its robust,
heavily ridged black bakelite cabinet
was manufactured during and after
World War II. Its cabinet design
remained the same, the only
alteration being in 1945 with its
change of waveband.

RIGHT

Misha Black's UAW78 Ekco of 1937
seems to hark back to Serge
Chermayeff's Ekco AC74. It is
relatively rare, especially in black
with chrome trim.

LEFT

The model AD76 is now recognized as the most technically accomplished Round Ekco. Produced in 1935 as the successor to Wells Coates' original design, it is recognizable by its distinct thick horizontal chrome bar, which was sprayed brown for the "walnut" version.

RIGHT

The Round Ekco was scaled down with the model AD36 of 1935. £3 3s cheaper and 25% smaller than its predecessor, it is instantly recognizable by its two downward-curved bars. This model was also available to special order in a number of non-standard colors such as yellow and blue that are now considered incredibly rare.

ABOVE

A beautifully thought-out finale to the Round Ekco story, the model A22 of 1945 was produced in "walnut" or black bakelite, with either a "florentine bronze" or chrome loud-speaker surround. Its circular perspex dial had a traveling cursor that followed around the circumference of the cabinet. It was the culmination of twelve years of production, fourteen years of thought and five quite different designs, and resulted in the most logically designed Ekco that brought together both great form and faultless function.

LEFT

The Philips model 634A is known affectionately as the Ovaltiney Set after its appearance in an English television commercial in the early Eighties. With its bakelite loudspeaker and dial surround, it is an excellent example of a decorative marriage between bakelite and wood.

ABOVE

This transportable midget known as the Sobelette was manufactured by Sobell Industries in 1949. Marketed in brown bakelite, it was also available in sprayed finishes.

RIGHT

This perspex sphere with transparent twirly decorations was manufactured by Champion in 1947 as the Venus. Available in a number of pastel colors, its design must surely be thought of as the pinnacle of kitsch. The sphere is often mistaken for bakelite, though perspex is a much later plastic.

LEFT

This miniature portable, manufactured in perspex by Pye as the model M78F, was ill-fated. Resembling almost exactly the Japanese flag, the symbol of Britain's former enemy, it was swiftly withdrawn. It is said that all remaining sets were slung on a bonfire, resulting today in a much-desired and rare set.

LEFT

An early use of bakelite for radios can be seen in this functional Philips 2511 of 1929 with a black metal frame with panels of Arbolite imitation wood.

ABOVE

Made from Arbolite, this early receiver was manufactured by Philips in Holland in 1930. It came with a separate heptagonal speaker that either sat on the top of the radio or could be placed well away from it.

ABOVE

This crystal set, manufactured from heat-formed imitation tortoiseshell celluloid, was produced in 1925 by Kenmac of London. Designed to sit on the bookshelf, it needed a pair of earphones and a long outdoor aerial for optimum performance.

ABOVE

This 2-tube German–British Burton receiver was probably the first Art Deco bakelite radio, produced in 1928–9. It required a separate loudspeaker.

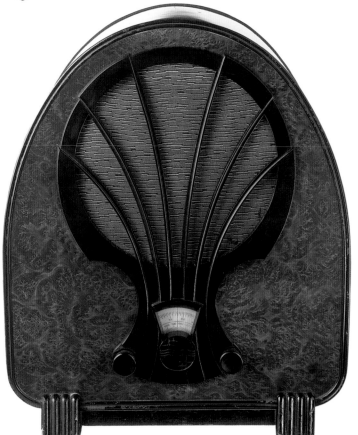

LEFT

This simulated walnut, Arbolite-cased Philips radio, known as the model 830A, was produced both in England and the Netherlands in 1932. It displays the typical sunburst speaker fret design synonymous with the Thirties.

AUSTRALIA

Not well known for its radio industry, Australia produced a number of stunning architecturally inspired radios such as the monumental AWA Radiolette, which when found in its soapy green variety epitomizes the color and feel of its age. The very small population of Australia resulted in a limited production run of each model, making today's examples, when found in good condition, rare and much sought.

RIGHT

Modeled on the Australian Wireless Association's office building in Sydney, this Radiolette is typical of the Empire State skyscraper style of the Thirties. With its deeply indented, bulbous ridges, it is particularly sought in this soapy green color.

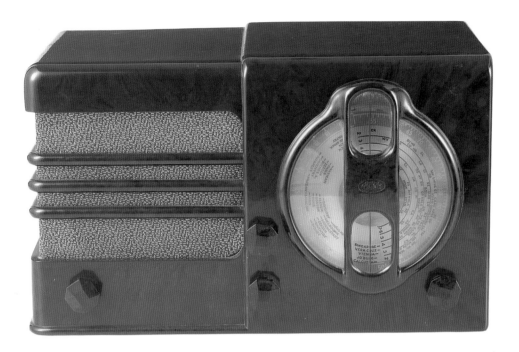

The organic-looking dial of this Astor from the mid-Thirties is a unique feature of this radio's design. The color of the dial changes from blue to yellow when the radio's frequency band is switched.

LEFT

The Healing Moderne radio of 1950 manufactured in either brown, white or occasionally green is often referred to as the "bathroom scales model." With its enormous dial and automobile-fender-grille front, it is surely one of the most overpowering designs ever produced.

ABOVE

Known as the Fisk Radiolette, its contrasting green fretwork grille and feet set against its lustrous black bakelite body makes the design of this dual-waveband radio stand out.

LEFT

Usually found bearing the AWA Radiolette insignia, this early General Electric of 1932 was the first bakelite radio manufactured in Australia. With its ornate, Art Nouveau cabinet and spiraling knobs, it must surely be one of the most exciting brown bakelite radios ever manufactured. Both extremely rare and much sought, specially in the near-mint condition of this example.

RIGHT

The Astor Football was produced in a multitude of pastel shades as well as the more often found brown. As with a majority of Australian radios, different dials were supplied, depending on which state you lived in. This one is for New South Wales.

FRANCE

With its own definitive style, France produced many uniquely shaped radios, including the series of three postwar mirrored-dial radios produced by the French–American company Sonora. At the leading edge of inventive, innovative design, they were based on the front grilles of flamboyant American automobiles. Not available in any great numbers, French radios are highly regarded in many countries, though they are not widely collected in Great Britain. The midget models are particularly sought, and in their colorful variants can exchange for large sums of money.

LEFT

This pillow-shaped radio, known as the Sonorette, was manufactured by a French–American company by the name of Sonora. In the late 1940s their radios were greatly influenced by the bulbous and curvaceous styling found on American automobiles. It was produced in a variety of colors, this green radio being among the scarcest.

Known affectionately as the Cadillac
due to its obvious bonnet-like
appearance, this American-designed
radio with its large mirrored dial and
chrome trim was manufactured by
Sonora in 1948 as the Excellence
model 301.

The link between the Sonorette and
the Cadillac. The Sonora 211 seems
to have mixed the two models
together, producing a rather quirky
hybrid that is now quite scarce.

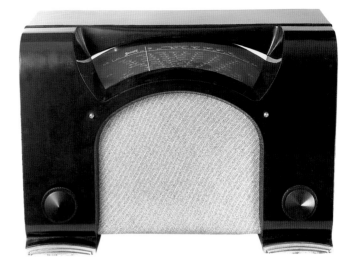

LEFT

With its unique illuminated glass pillars, this robust Sonora of 1935 is one of the closest relatives of the larger, more often seen Jukebox. It is quite scarce, especially as the German occupation required all radio receivers to be handed in.

ABOVE

A large Sonora that looks as though it belongs right in the heart of Paris! Its curved dial, large speaker expanse, and chrome pediments make this radio stand out as a bold example of Odeon style and design.

RIGHT

This sharply molded Ducretet from 1934 is unusual and pleasing in its cubist shape. It utilizes the company's tuning fork motif, turning it into a striking speaker-grille design.

RIGHT

France's first bakelite radio, the Radialva, was produced at the start of the Thirties. Its bold, geometric cabinet design and delicately decorated grille cloth hide the simplicity of the radio's construction.

BELOW

The Super Groom, model 41, manufactured by Radialva in the late Thirties. It displays three wavebands, GO, PO and OC, or Long, Medium and Short Wave, on its large, oval-shaped glass dial.

ITALY

A limited number of receivers were manufactured in Italy during the pre- and postwar years due to the country's deep financial recession. This resulted in a none-too-prolific radio industry making the majority of their radios in wood. When bakelite was used, the inventiveness and strength of design was outstanding, as with Castiglioni's sculptural Phonola of 1939. Today Italy is a far more prosperous nation, and with past history being pursued with great fervor radio prices are exceptionally high, where sometimes a good example cannot be purchased for "love nor money!"

ABOVE

Although not prolific in their manufacture, Italian radios exude their own certain style of design. The four wavebands of this robust Phonola 557 or 1937 light up individually when a specific column is selected for tuning.

RIGHT

Designed by Castiglioni, this Phonola of 1939 is the only radio to be exhibited in New York's Museum of Modern Art. Curvaceous and organic in appearance, it could either be wall-mounted or laid flat. It was manufactured in a variety of colors, this dark green example being one of the rarest.

BELOW

Produced in 1939, this Radiomarelli (known affectionately as the Fido) takes its styling directly from compact American prewar radios such as the Emerson and RCA midgets.

SPAIN

Civil war and poverty laid to rest any plans of bakelite radio production in Spain, which only got going in the postwar Forties when a number of large foreign manufacturers set up factories in both Madrid and Barcelona producing Spanish variants of current models. When homegrown designers were finally allowed a free hand in the mid-Fifties, a number of striking shapes such as the chrome-covered Artes appeared. Never produced in large quantities, the quirkier models are now much prized.

RIGHT

Telefunken's dramatic model 1065 seems to take its inspiration from the orchestral, wooden harp. Its brown bakelite case is ridged and indented with precise gold painted lines.

This Iberia of 1953 takes its dashboard, bullet-nose styling from its North American automotive counterpart, the Crosley. Its case is composed of white sprayed, hardened paint onto bakelite.

ABOVE

Philips among other companies set up a factory in Barcelona producing this highly stylized miniature, sunburst design with its model BE 312U of 1952.

RIGHT

The German company, Telefunken set up a factory in Madrid where it produced a multitude of radios both pre- and postwar. This striking, bulbous white example, the model U1515, was manufactured in the mid-Fifties.

This factory-painted deep red Suministrado was marketed in kit form for the home enthusiast to assemble during the mid-1940s.

Britain was not the only country to produce a circular radio – Aeesa also manufactured a miniature version in 1955 known as the Estrella Polar. Its curvaceous flayed out legs supporting its rotund body give this small midget radio an immense amount of charm and character. It looks almost exactly like the Wells Coates-designed Round Ekco sets of the mid-Thirties.

An automobile influence is obvious in this Artes AR3 receiver in sprayed bakelite, but the chrome horse-shoe and sunray speaker fret seem to derive from a Philips design of 1932.

GERMANY

The first country to realize the enormous potential of bakelite in the radio industry, Germany started manufacturing bakelite sets as early as 1926. Telefunken and Mende produced a large array of bold monolithic radios of a striking form that are relatively scarce today because of the standardization and subsequent confiscation of any set that did not receive the single waveband propaganda speeches of Hitler's Germany. These mass-produced, economically constructed sets with their Nazi regalia were manufactured in large factories such as Siemens and Telefunken, and are fairly common today.

BELOW

This Loëwe set was innovative in many ways. It was one of the first bakelite radios and the double tube was the first example of an integrated circuit, containing almost all the components needed for a radio, apart from a tuning device and an on–off switch. This version was marketed in the UK.

The Nora Sonnetblume of 1930, literally translated "The Sunflower," is a spectacular example of the designer's hand running wild. The company was owned by a Jewish family by the name of Aaron, an obvious reversal of Nora.

RIGHT

The marbleized Beetle trim on this rare 1934 Lumaphon is very brittle. It makes finding this austere radio, reminiscent of the Nazi style of architecture, hard to find in a completely intact condition.

ABOVE

Saba produced a substantial array of
large radios during the early Thirties.
They are distinctive for their
fretwork-type treatment of the
speaker grille, as in this magnificent
model 321GL of 1934.

ABOVE

The Deutscher Kleinempfänger, or "Peoples Set," was produced by a conglomerate of radio manufacturers in 1938. With its centralized Eagle-clasping-Swastika emblem, it was constructed to receive only stations within the Fatherland, and is now known after Hitler's Propaganda Minister as the Goebbels Schnauze ("Big Mouth"). Note that with this particular example, an attempt has been made to scratch out the Swastika.

LEFT

In 1935 the Volksempfänger VE301GW was the first radio churned out for the masses. It displays an eagle emblem synonymous with the symbols of the Third Reich.

The modernist, upmarket "Party" version of the Kleinempfänger, the VE301 was produced in 1938, the year of the Anschluss. As Austria and Germany joined so did the station names on the radio's dial. The swastika on this set is clearly visible.

Mende manufactured a number of imposing tombstone-shaped radios in the early Thirties. This example, the model 148, was produced both in bakelite and in wood. Nowadays these beautifully styled radios are highly prized and much sought.

This 1938 DAF bakelite tuner–amplifier was used in German factories, connected to public address systems to provide music to aid productivity. They also relayed Nazi propaganda – particularly Hitler's speeches, which were compulsory listening during the war.

ABOVE

A bold Saba radio from 1933
showing an unusual deeply
indented, faceted front and
geometrical latticed grille.

CZECHOSLOVAKIA

A number of major factories were based in Czechoslovakia in the years preceding World War II. As well as their own Europe-wide models a number of variants were manufactured, such as the charming Philips Butterfly of the mid-Thirties. Following the Occupation, bakelite radio production all but ceased. As a result many examples have been well used and are therefore in poor condition.

LEFT

This rather small Philips radio was locally manufactured in the early Thirties. Known as the Butterfly for obvious reasons, the model 964AS is considered rare today.

RIGHT

The streamlined Tesla Talisman, although designed in the Thirties, was only put into production in the late Forties. Highly desirable in Iron Curtain days, it has now turned up in great numbers, causing its value to plummet for the time being.

GAD SASSOWER

BIBLIOGRAPHY

Aitken, Professor Hugh, *Syntony and Spark*, USA: 1977.

Arts Council, *The Thirties – British Art and Design before the War*, London: 1979.

Baker, W. J., *A History of the Marconi Company*, London: Methuen, 1970.

Banham, Reyner, *Theory and Design in the first Machine Age*.

Barthes, Roland, *Mythologies*, Vintage 1993.

BBC Handbooks and Yearbooks, 1928 onward, London.

Bergonzi, Benet, *Old Gramophones*, Shire: 1991.

Biraud, Guy, *Les Radio Philips de Collection*.

Biraud, Guy, *Guide du Collectionneur* and *La Restauration et la Conservation*, France: 1987.

Boselli, Primo, *Il Museo Della Radio*, Florence: Edizioni Medicea, 1989.

Briggs, Asa, *The History of Broadcasting in the UK*, Oxford University Press: 1961–79.

Briggs, Susan, *Those Radio Times*, London: Weidenfield and Nicholson, 1981.

Britt, David (Ed.), *Modern Art Impressionism to Post-Modernism*, London: Thames and Hudson, 1989.

Bunis, Marty and Sue, *Collector's Guide to Antique Radios*, USA: Collector Books (Schroeder), 1991.

Burrows, Arthur, *The Story of Broadcasting*, 3 vols, London: Cassell, 1924.

Cabinet Maker, The: London 1930–3.

Carrington, Noel, *British Achievement in Design*, London: 1946.

Chew, V. K., *Talking Machines*, London: Science Museum, 1967.

Claricoats, John, *The World At Their Fingertips*, London: RSGB, 1967.

Collins, Michael, *Towards Post-Modernism*, London: British Museum, 1987.

Constable, Anthony, *Early Wireless*, London: Midas, 1980.

Cook, Patrick and **Slessor**, Catherine, *Bakelite, an Illustrated Guide to Collectible Bakelite Objects*, London: Quintet/The Apple Press, 1992.

Dalton, W. M., *The Story of Radio*, 2 vols, Inst. of Physics, 1975.

De Vries, Leonard, *Victorian Inventions*, London: John Murray, 1971.

DiNoto, Andrea, *Art Plastic, Design for Living*, Abbeyville, 1984.

Douglas, Alan, *Radio Manufacturers*, several volumes, New York: 1988 onwards.

Eckersley, Peter, *The Power Behind the Microphone*, London: Jonathan Cape, 1941.

Eisler, Paul, *My Life with the Printed Circuit*, AUP, 1989.

Ernst Erb, *Radio von Gestern*, Switzerland: M+K Computer Verlag AG.

Forester, Tom (Ed.), *Microelectronics Revolution*, Oxford: Blackwell, 1980.

Forty, Adrian, *Objects of Desire*, London: Thames and Hudson, 1986.

Forty, Adrian, "Wireless Style", *Architectural Association Quarterly*, vol. 4, 1972.

Freud, Sigmund, *The Interpretation of Dreams*, Standard Edition, London: Pelican, 1976.

Freud, Sigmund, *The Psychopathology of Everyday Life*, London: Pelican, 1976.

Grinder, Robert and **Fathauer**, George, *Radio Directory*, USA: 1986.

Ham, Ron and **Rudram**, David, *History of Communications*, Amberley Industrial History Museum.

Hawes, Robert (Ed.) *Bulletin of the British Vintage Wireless Society*, London: 1982–1994.

Hawes, Robert, *Radio Art*, London: Greenwood, 1991.

Harmsworth Wireless Encyclopaedia, c1923.

Heskett, John, *Industrial Design*, London: Thames and Hudson, 1980.

Humphries, Lund, *Eye for Industry*, London: Royal Society for Arts, 1986.

International Design Yearbooks, London: Thames and Hudson.

Jensen, Peter R., *In Marconi's Footsteps – Early Radio*, Australia: Kangaroo Press, 1994.

Jessop, George, *The Bright Sparks of Radio*, London: RSGB, 1990.

Johnson, David and Betty, *Antique Radios and Guide to Old Radios*, USA: Wallace-Homestead, 1982 and 1989.

Jung, Carl, *Man and his Symbols*, London: Aldus/Jupiter, 1964/74.

Katz, Sylvia, *Classic Plastics*, London: Thames and Hudson, 1984.

Katz, Sylvia (Ed.), *The Plastics Age*, London: Victoria and Albert Museum, 1990.

Katz, Sylvia, *Early Plastics*, London: Octopus, 1974.

Klein, Dan, *Art Deco*, London: Octopus, 1974.

L'e Turner, Gerard, *19th Century Scientific Instruments*, London: Sothebys/Philip Wilson, 1983.

Lethaby, W., *Form in Civilisation*, London: 1922.

Littmann, Frederic, "The evolution of the wireless receiver," *Design for Today* magazine, March 1936.

Lodge, Sir Oliver, *Talks about Wireless*, London: Cassell, 1925.

Long, Joan, *A First Class Job*, Murphy, 1985.

MacKenzie, Donald and **Wajcman**, Judy: *The Industrial Revolution in the Home*, Oxford University Press, 1985.

Miller, Charles, *Practical handbook of valve radio repair*, London: Newnes, 1982.

O'Dea, *Handbook: Radio Communications Collections*, London: Science Museum, 1934.

O'Neill, Amanda (Ed.), *Introduction to the Decorative Arts*, London: Tiger, 1990.

Packard, Vance, *The Hidden Persuaders*, London: Longmans, 1957.

Paul, Floyd, *Radio horn speaker encyclopaedia*, USA: 1987.

Pevsner, Nikolaus, *Enquiry into the state of Industrial Art in England*, Cambridge University Press: 1937.

Pevsner, Nikolaus, *Pioneers of Modern Design*, New York Museum of Modern Art, 1949.

Pevsner, Nikolaus, "The Radio Cabinet," *Architectural Review*, London, May 1940.

Povey, P. J. and Earl, R. A. J., *Vintage Telephones*. London: Peregrinus/London Science Museum, IEE, 1988.

Proudfoot, Christopher, *Collecting Phonographs and Gramophones*, London: Christies/Vista, 1980.

Rowlands, Dr Peter and **Wilson**, Dr Patrick, *Oliver Lodge and the Invention of Radio*, Liverpool: P.D. Publications, 1994.

Royal Academy, *50 Years of the Bauhaus*, London: 1968.

Russell, Gordon: *The Designer's Trade*, London: 1968. Also articles in *Design for Today* magazine, 1933.

Sideli, John, *Classic Plastic Radios*, New York: Dutton, 1990.

Soresini, Franco, *La Radio*, Milan: BE-MA Editrice, 1988.

Sparke, Penny, *An Introduction to Design and Culture in the Twentieth Century*, London: Allen and Unwin, 1986.

Sparke, Penny, *The Plastics Age, from Modernity to Post-Modernity*, London: Victoria and Albert Museum, 1990.

Stokes, John, *The Golden Age of Radio in the Home*, New Zealand, 1986, and further volume.

Stokes, John, *70 Years of Radio tubes*.

Sturmey, S. G., *The Economic Development of Radio*, London: Duckworth, 1958.

Tyne, Gerald, *Saga of the Vacuum Tube*, USA.

Van de Lemme, Arie, *A Guide to Art Deco Style*, Quintet/Magna Books: 1992.

Wander, Tim, *2MT Writtle: The Birth of British Broadcasting*, Chelmsford: Capella, 1988.

Ward, Peter, *Kitsch in Sync – a Consumer's Guide to Bad Taste*, London: Plexus, 1991.

Whiteley, Nigel, "Toward a Throwaway Culture," *Oxford Art Journal*, vol. 10, 1987.

Wyborn, E. J. and **Landauer**, W., *Plastics and Trend*, London: Summer 1936.

INDEX

PICTURE CREDITS

Supplied by Robert Hawes: p11 (GEC Marconi); pp12, 13, 14; pp16, 17 (GEC Marconi); p18; p25 *top left* (Design Council, London); p29 (*top* GEC Marconi); p35 *top* (Design Council, London); p37 (Colección Julia); p41 (*top* J. Juliá, Spain); pp49, 50; p51 (*top left* Bharat Goswami), (*top right* John Sideli), (*bottom* Design Museum, London); p53; p57 (J. Juliá, Spain); p94 *center*; p106 *bottom left*; p119 *center and bottom* (J. Juliá, Spain); p124 *center* Rudiger Walz.

Supplied by Gad Sassower: pp30, 36, 45, 46, 64 *left*, 67 *left*; 79 *top right* (James Meehan); 82 *center* (James Meehan); 117, 118, 119 *top* (J. Juliá, Spain), 124 *bottom*.